The Future of Science

Contributors

JOHN C. ECCLES

LANGDON GILKEY

POLYKARP KUSCH

GLENN SEABORG

THE FUTURE OF SCIENCE

1975 NOBEL CONFERENCE

organized by

Gustavus Adolphus College
St. Peter, Minnesota

edited by

TIMOTHY C. L. ROBINSON
Gustavus Adolphus College

A WILEY-INTERSCIENCE PUBLICATION

JOHN WILEY & SONS, New York • London • Sydney • Toronto

Library of Congress Cataloging in Publication Data

Nobel Conference, 11th, Gustavus Adolphus College, 1975.
   The future of science.

   "A Wiley-Interscience publication."
   Includes bibliographical references.
   1. Science—Social aspects—Congresses. I. Eccles,
John Carew, Sir. II. Robinson, Timothy C.L., 1943–
III. Gustavus Adolphus College, St. Peter, Minn.
IV. Title.
Q175.4.N6   1975   301.24'3   76-49607
ISBN 0-471-01524-5

1975 Nobel Conference Participants

NOBEL LAUREATES

Christian Anfinsen—Chemistry, 1972
National Institutes of Health

George Beadle—Medicine, 1958
University of Chicago

Hans Bethe—Physics, 1967
Cornell University

Felix Bloch—Physics, 1952
Stanford University

Walter Brattain—Physics, 1956
Whitman College

Leon Cooper—Physics, 1972
Brown University

André Cournand—Medicine, 1956
Columbia University

Christian de Duve—Medicine, 1974
Rockefeller University

Sir John C. Eccles—Medicine, 1963
State University of New York at Buffalo

Gerald Edelman—Medicine, 1972
Rockefeller University

Robert Hofstadter—Physics, 1961
Stanford University

Charles Huggins—Medicine, 1966
University of Chicago

Polykarp Kusch—Physics, 1955
University of Texas at Dallas

Simon Kuznets—Economics, 1971
Harvard University

Willis Lamb, Jr.—Physics, 1955
University of Arizona

Willard Libby—Chemistry, 1960
University of California at Los Angeles

Fritz Lipmann—Medicine, 1953
Rockefeller University

Robert Mulliken—Chemistry, 1966
University of Chicago

Lars Onsager—Chemistry, 1968
University of Miami

Julian Schwinger—Physics, 1965
University of California at Los Angeles

Glenn Seaborg—Chemistry, 1951
University of California at Berkeley

Emilio Segrè—Physics, 1959
University of California at Berkeley

William Shockley—Physics, 1956
Stanford University

Ulf von Euler—Medicine, 1970
Royal Caroline Institute, Stockholm

Ernest Walton—Physics, 1951
University of Dublin

Thomas Weller—Medicine, 1954
Harvard University

Chen Ning Yang—Physics, 1957
State University of New York at Stony Brook

THEOLOGIANS

Ian Barbour
Carleton College

John Cobb, Jr.
*School of Theology
Claremont, California*

William Dean
Gustavus Adolphus College

Langdon Gilkey
University of Chicago

Van Austin Harvey
University of Pennsylvania

Hans Schwarz
*Lutheran Theological Seminary
Columbus, Ohio*

The Contributors

Sir John Eccles was born in Melbourne, Australia, in 1903. After receiving M.B. and B.S. degrees from Melbourne University in 1925, he attended Oxford University on a Rhodes Scholarship, studying physiology under Sir Charles Sherrington and gaining a Ph.D. in 1929.

He remained in England as a tutor and lecturer in physiology at Magdelen College until 1937, when he returned home to become Director of the Kanematsu Memorial Institute of Pathology in Sydney, where he served until 1943. He served as Professor of Physiology at the University of Otago Medical School in Dunedin, New Zealand from 1944 to 1951. In 1952 he established his laboratory at the John Curtin School of Medical Research of the Australian National University in Canberra, where he performed much of the work that earned him a Nobel Prize in Medicine in 1963 with Alan Hodgkin and Andrew Huxley. Since 1966 he has been associated with the American Medical Association's Institute for Biomedical Research in Chicago and served as Distinguished Professor of Physiology and Medicine at the State University of New York at Buffalo.

During his long career, Dr. Eccles made significant contributions to nearly every aspect of the field of neurophysiology. As a pioneer in the use of microelectrode recording techniques, his early work resulted in better understanding of the complex processes involved in synaptic transmission. His more recent investigations of the cerebellum formed the basis for his exceptionally clear statement of the function and information processing capabilities of this major brain structure.

Unlike many of his colleagues, Dr. Eccles has always maintained an interest in the philosophical aspects of the relation between the mind of man and the neuronal activity of his brain, an interest first nurtured by his association with Sir Charles Sherrington, and later by the philosopher Karl Popper.

A writer with a lucid and penetrating style, Dr. Eccles has published over 350 scholarly papers and received numerous honors from scientific organizations. He received knighthood from Queen Elizabeth II in 1958. He is a member of the Nobel Conference Advisory Committee and has spoken at two previous Nobel Conferences.

Langdon Gilkey, eminent theologian and author, was born in Chicago in 1919. He was educated at the Asheville School for Boys in Asheville, North Carolina and at Harvard University, where he earned a B.A. degree in 1940.

While teaching English at Yenching University near Peking in 1941, he was captured by invading Japanese soldiers and interned for the duration of the war in a concentration camp in Shantung province. His observations of the hardships endured during this imprisonment are recorded in his book, *Shantung Compound.*

On returning from China, he gained a M.A. degree from Union Theological Seminary in 1949 and taught at Vassar before earning a Ph.D. in religion from Columbia University and moving to the Vanderbilt University School of Divinity in 1954. Since 1963 he has been Professor of Theology at the University of Chicago Divinity School.

Dr. Gilkey was a Fulbright Scholar in England in 1950 and a Guggenheim Fellow in 1960 and 1965. He is the author of *Maker of Heaven and Earth, How the Church Can Minister to the World Without Losing Itself,* and *Naming the Whirlwind: The Renewal of God-Language.* His ideas on the relationship between science and theology and the role of each in modern society can be found in his book, *Religion and the Scientific Future.*

Polykarp Kusch was born in Blankenburg, Germany, in 1911. He and his family moved to Cleveland, Ohio shortly thereafter, where he attended public high school. He holds the degrees of Bachelor of Science from the Case Institute of Technology, and Master of Science and Ph.D. in physics from the University of Illinois.

He began his career at the University of Minnesota, moving to New York City after accepting an instructorship in the physics department at Columbia University in 1936. He left there to work on radar research in 1941 at the Westinghouse and Bell Telephone Laboratories, but rejoined the Columbia faculty in 1946, serving as Chairman of the Physics Department, Director of the Columbia Radiation Laboratories, and eventually, Vice President for Academic Affairs and Provost of the University. In 1972 he became Professor of Physics at the University of Texas, Dallas, where he was named Eugene McDermott Professor in 1975.

His research has been primarily in the area of the application of quantum mechanics to atomic physics. He was awarded the Nobel Prize in physics in 1955 with Willis Lamb for his precision determination of the magnetic moment of the electron. This work subsequently led to theoretical statements that form the basis for quantum electrodynamics.

Although he is widely known throughout the scientific community for his precise research, he has also demonstrated a continuing interest in science education. In his witty and enthusiastic lectures to both beginning and graduate students he has long stressed the purpose and cultural impact of science in addition to the facts of the discipline.

Dr. Kusch has been a participant in two previous Nobel Conferences and is a member of the Nobel Conference Advisory Board.

Glenn T. Seaborg was born in Ishpeming, Michigan in 1912. After attending public high school in Los Angeles, he received an A.B. degree in chemistry from the University of California, Los Angeles, and in 1937 a Ph.D. in nuclear chemistry from the University of California, Berkeley.

Dr. Seaborg joined the faculty at the University of California, Berkeley in 1939 after serving as the research assistant to Gilbert Newton Lewis, the chemist. With the exception of leaves to head the Metallurgy Laboratory of the University of Chicago during World War II and to serve as Chairman of the Atomic Energy Commission from 1961 to 1971, his entire academic career has been spent at Berkeley, serving the university as a teacher, Director of Nuclear Chemical Research, Associate Director of the Lawrence Berkeley Laboratory, and as Chancellor from 1958 to 1961.

He is best known for his work on the chemistry of transuranium elements. Beginning in 1940 when he co-discovered the element plutonium, his and his colleagues' research has led to the discovery of nine other elements and over 100 isotopes. As the author of the actinide concept of heavy element structure, Dr. Seaborg's theoretical work also permitted the classification of these newly discovered heavy elements in the periodic table. He received the Nobel Prize in chemistry with E.M. McMillan in 1951.

Throughout his career, Dr. Seaborg has been a forceful advocate of the peaceful use of nuclear energy. He was head of the United States delegation that signed the "Memorandum on Cooperation in the Field of Utilization of Atomic Energy for Peaceful Purposes" in the USSR in 1963, and was a member of the United States delegation to Moscow for the signing of the Nuclear Test Ban Treaty in 1971.

Dr. Seaborg is currently the President of the American Chemical Society and is a member of the governing boards of numerous organizations, including the American–Scandinavian Foundation. He has been a member of the Nobel Conference Advisory Committee since 1965 and a participant in two previous Nobel Conferences.

Preface

The association between the Nobel Foundation and Gustavus Adolphus College began in 1963 when twenty-six Nobel laureates gathered to dedicate the college's Nobel Hall of Science as the first American memorial to Alfred Nobel. Two years later the Nobel Foundation gave its endorsement to organize a conference in which scientists and theologians could meet to discuss the leading topics in science for the benefit of an audience of intelligent laymen.

This year's topic, "The Future of Science" was suggested by Dr. Glenn Seaborg. Perhaps remembering the initial gathering of laureates, at which he was a keynote speaker, he suggested that a similar meeting of distinguished scientists at the Nobel Conference might provide an effective forum in which the role of science in society might be evaluated and in which the fears and aspirations of scientists for the coming years could be meaningfully expressed.

On October 1 and 2, 1975, twenty-seven Nobel laureates and six distinguished theologians met before some 4000 theologians, scientists, and science students to discuss the future of science. The audience could hardly have predicted the variety of views they were about to hear.

Dr. Glenn Seaborg, while noting that the problems of over-population, dwindling resources, and unequal distribution of wealth pose a very real threat to man's survival, remained optimistic that science and technology will play a major role in finding solutions to these problems, but he also warned that immediate steps must be taken to replenish the "knowledge capital" of science and that reductions in our standard of living may be necessary to solve these problems on a global scale.

Dr. Polykarp Kusch took a personal view of science. Recalling events in his career, he emphasized the aesthetic pleasures of discovery as the motivating force of pure science and considered what similar conferences in the past may have predicted. He expressed concern, however, that the purpose and spirit of science have not been adequately communicated to the society on whose financial support and understanding the future development of science depends.

Sir John Eccles explored the frontiers of science by restating the mind-body question in a new conceptual framework. He posed a challenge to the methods of science by calling for an enlarged definition of science and insisting that science should begin to question matters previously left to religion and philosophy.

Dr. Langdon Gilkey echoed the sentiments of the detractors of science. He likened science to the religion of the Middle Ages, a "queen" whose influence pervades every aspect of modern culture. He called on science to assume a larger responsibility for its discoveries and to understand better the power of its knowledge in influencing the society we live in.

Following the addresses, five discussion panels were formed consisting of laureates and at least one theologian. The ensuing discussion was for the most part spirited, sometimes highly technical, and occasionally argumentative. Obviously, not all 20 hours of discussion can be related in a meaningful fashion, but the sample presented reflects an attempt to capture both the substance and the spirit of these stimulating sessions.

TIMOTHY C. L. ROBINSON

St. Peter, Minnesota
November 1976

Acknowledgments

The 1975 Nobel Conference was made possible by a grant from members of Aid Association for Lutherans (AAL). In honoring its longtime president, Walter L. Rugland, AAL made provision to allow theologians and students from all Lutheran colleges and seminaries to meet at the Rugland Assembly held in conjunction with the conference—a fitting tribute to a man of both conviction and wisdom.

Appreciation also goes to the Nobel Foundation, to Baron Stig Ramel, President of the Nobel foundation, and to the foundation's envoy to the Nobel Conference, Dr. Ulf von Euler, himself a Nobel laureate and past chairman of the board.

As editor, I wish to express my appreciation to Elaine Brostrom for her remarkable transcriptions of the discussion sessions, to Bev Lee, Carole Mataya, Stephanie Kendall, and Bryan Klingberg, all invaluable in the preparation of the manuscript, and to my wife Sharon, the final arbiter of English language usage, who suffered the burden of the proofreading, among other things.

T. R.

Tribute to
Dr. Edward Lawrie Tatum

This volume of the Nobel Conference Lectures at Gustavus Adolphus College honors the memory of Edward Lawrie Tatum whose amazingly creative career ended with his death on November 5, 1975. His health had prevented him from attending and participating in this year's conference on October 1 and 2. He was intimately and directly involved with these Nobel conferences from the beginning. As one of the twenty-six Nobel laureates who attended the dedication of the Nobel Hall of Science in 1963, he had served continuously since that time as a member of the Nobel Advisory Committee planning the conferences, and he was one of the lecturers at the first Nobel Conference on "Genetics and the Future of Man." He had much to do with the continuing high quality of these lectures through his suggestions for themes and lecturers over the years.

But it is, of course, as a research scientist of the very highest order that he will be remembered. His unique achievements were recognized when he became a co-recipient with George W. Beadle and Joshua Lederberg of the 1958 Nobel prize in medicine and physiology for "discovering that genes act by regulating specific chemical processes." Through extensive experimentation with the mold *Neurospora crassa* (with Beadle) and the common bacterium *Escherichia coli* (with Lederberg) Tatum helped establish a direct and specific correlation between genes and chemical reactions. By using X-rays to produce mutations, some of which were deficient in ways that prevented growth, and finding the nature of that deficiency in each case, they were able to substantiate their hypothesis that genes control biochemical reactions in cells.

Their work opened the door to a very large number of studies on the chemical nature of the gene and the chemical process whereby its information is converted into the structure of proteins. A part of this exploration has involved the unraveling of the genetic compound deoxyribonucleic acid (DNA) and the way its subunits form a code. A number of discoveries building on the Tatum-Beadle–Lederberg achievements have earned Nobel prizes in subsequent years. By demonstrating that microorganisms could be used for genetic investigations, they were able to greatly accelerate the experimental process. Another side effect has been the possibility of making antibiotics much more efficient and adaptable.

Dr. Tatum held degrees (A.B., M.S., Ph.D.) from the University of Wisconsin. His major work was done at Stanford University (1937–1945 and 1948–1957), Yale University (1945–1948) and Rockefeller University (1957–1975).

He was a cordial and inviting person, with broad human and cultural interests. He was an enthusiastic musician and was greatly concerned about the physical and political environment in which responsible scientific inquiry could be conducted.

EDGAR M. CARLSON
President Emeritus
Gustavus Adolphus College

Contents

The Future of Science   **105**

LANGDON GILKEY

    Discussion: General Comments
                    Responsibility and Science

The Future of Science

New Signposts for Science

by

GLENN T. SEABORG

We are here to discuss the future of science. To do so is nothing less than to probe the very future of mankind, for it is the course of science — reflecting man's ability to increase his knowledge and apply it wisely, together with his spirit for creative and constructive change — that will determine that future.

Because we are a conference of both scientists and theologians, of both specialists and generalists, but above all, of human beings vitally concerned with the human condition as a whole and the direction in which man is moving, let me briefly state my position about the spirit of man. Then I will move on to focus on the role of science.

There are, unfortunately, many who see the spirit of man in a state of decline. As symptoms of that decline they usually cite such factors as an excessively materialistic outlook on the part of the advanced nations, a degrading of the environment and a wasting of nonrenewable resources, a rising feeling of nationalism combined with a hardening attitude toward development, and a breaking down of social systems through all levels of government and institutions right down to the family unit itself. To some extent there is evidence to support all these claims. The media, highlighting and emphasizing all the symptoms of the world's problems, lend strong emotional support to this picture of decay and doom.

But without denying for a moment the seriousness of the world's problems and the challenges they pose, I contend that there is increasing evidence to show that men and women all over the world are reacting positively and strongly in response to these challenges. The available but usually unheralded facts confirm that never before has there been such concern, thought *and action* on the part of so many individuals and institutions — national and international — over the condition of man and the world as it exists today. There is no guarantee that all these responses will be correct or productive. As a working out of our evolutionary process (about which I have more to say later) many such responses will fail, and some may even prove counterproductive. Those that do prove to be so will most certainly take their toll, much as do all harmful mutations in the natural evolutionary process, but I firmly believe that with mankind the successful responses will prevail, moving us to a higher plane. Contrary to what some skeptics believe, neither our technologies nor our institutions will become the tar pits of humanity. We will survive and advance by changing them to meet our higher needs, rather than decline by adapting regressively to the poorer conditions and environments they might create as they become outmoded and harmful.

The fact that we are already in the process of doing this — which accounts for so much of the ferment and agitation visible today — is the reason I remain an optimist concerning the spirit of man. Now, let me turn to his science and outline for you some of the directions in which I see it moving. Here again, I am an optimist, but, as you will see, perhaps a cautious one.

Both science and technology will undergo a great many changes in the years ahead. In general, I think we will witness a brief period of conservatism for science — one reflected in the public support of science and perhaps within the science community itself. A more cautious, make-haste-slowly attitude will prevail for a while. We are already into that period.

I think we will see a continuing shift of interest toward the life sciences — particularly toward many facets of biology and biochemistry — and toward the social and behavioral sciences. This is already under way. Perhaps most noticeable will be a renaissance in engineering, as a number of new ideas about how we should live in the future begin to take hold and must be transformed into realities. This is not evident yet.

If we talk about the economic expansion of science in the future, I think we must come to terms with the fact that the rate of its growth will slow down. This, however, should be put in perspective, particularly in perspective with the last 25 to 30 years, which must be recognized as a period of exceptional growth. During the peak years of this period support of science increased at an average rate of 15% annually. Of course, this was a time of unusual economic growth not only for science and not only in the United States. As Lester Brown pointed out at the Second General Assembly of the World Future Society last June, the global economic output tripled between 1950 and 1975. We cannot expect the last quarter of this century to be an extrapolation of that trend, and in many respects we would not want it to be. Instead, a period of slower growth, one of maturation and transformation, will take place, all of which ultimately will prove more beneficial.

If we look directly at science, we can see many reasons why its growth will follow a similar course. In fact, science and technology are already on such a course and are now beginning to be subjected to a number of new influences and pressures. Let me mention a few of them.

First, there is the economic squeeze on the funding of research and development. This has several effects. In a country that historically has been rather pragmatic about its support of science, there is now, understandably, mounting pressure on science and technology for economic payoffs. Since the late 1960s there has been a slight shift of support from military- and space-related research to research oriented toward civilian needs. With economic conditions focusing greater attention than ever on such needs, significant results are now being expected, and it is difficult for the public to understand why science and technology cannot produce miracles on demand as they did in the Manhattan and Apollo projects. It does little good for scientists and engineers to respond that the nature of the problems are different, that they do not lend themselves to the same kind of solution, that the controls are lacking, and even that sometimes there is no common agreement on the goals. The public has too long been conditioned to the idea of crash programs and to the notion that the right combination of money and manpower can solve any problem within a set period of time. The concept that we live within what might be considered an ecology of problems — dynamically related and subject in any time frame less to complete solution than to amelioration and change — is just beginning to dawn on many people, including many in positions of authority and important decision making.

One result of the emphasis on quick solutions is that as the general economic situation tightens and as attention to priorities increases, the division of those funds that are allotted to research and development may be made too heavily in favor of applied and developmental work. This, of course, is not a new situation. There always has been difficulty in balancing basic and applied research. Even Louis Pasteur was torn between the need to advance the practical application of science and that of advancing the basic knowledge that underlies all progress. This was quite evident in his speech of 1854, in which he proudly announced the establishment, by imperial decree, of a new university degree, under the title of "Certificate of Ability in Applied Science." But then, after praising this degree as one that would certify scientists to enter industry and stressing the government's attachment to "the spread of applied knowledge," he launched into an impassioned defense of theoretical knowledge and its basis as "the mother of practice" that "alone gives rise to and develops the spirit of invention." As a caution to his fellow scientists, lest they let the pragmatic influence get too strong, he added: "It will be especially up to us not to share the opinions of those narrow minds who disdain, in the sciences, all that has no immediate application" (1).

Another important influence on the future of science is the other side of the cost/benefit coin. Although society is interested in greater and more immediate benefits from its science and technology, it is also more concerned in these times with their growing price tag, and there is reason for this concern. Today's methods and tools for conducting scientific research in most disciplines are more capital intensive than ever, and the cost of equipment and manpower has risen at a rate faster than the general rate of inflation. For example, in the field of physics — admittedly a costly discipline — the cost of instrumentation between 1945 and 1975 rose between 175 and 200% as compared with the average rise in the cost of living index of 75% for the same period (2). In addition to the increasing cost of basic and applied research, the overruns in developmental work are becoming notorious and giving rise to second thoughts about many important projects.

As a result of all this, we are going to see tighter budgeting and closer overseeing of all science and technology programs, with a greater demand for accountability. Depending, of course, on the attitudes and wisdom of those who have the ultimate control over the support of these programs, the results could have any number of effects. They could reduce waste and sharpen the focus of our objectives in a beneficial way. On the other hand, they could have the conservative effect of giving the major support to the most conventional type of research and to only the established researchers, tending to narrow the scope of work detrimentally and severely limiting efforts in new and potentially significant directions. But most certainly these tighter times will force many scientists to become more publicly responsive and articulate about their work and its ultimate contributions to society.

This brings up another kind of accountability — aside from the economic aspect — which will have a great bearing on the future of science and technology. It is the kind of accountability that concerns itself with the risk/benefit rather than the cost/benefit ratio. Its focus is not on the question of whether the work is worth doing, but on whether its potential harmful impact may outweigh any good it could do, on whether the research or project should be initiated at all. This influence on thinking related to research and development, which has risen rapidly over the past half-dozen years and is being institutionalized by such requirements as environmental impact studies and technology assessments, is already a major factor in determining the future of science. It is affecting work on energy resources and technologies, biological research, aircraft development, advances in communications and the use of the computer, and even research in the social sciences and education.

Almost every Federal agency involved in the support of research and development in a scientific or technological field can attest to the growing effect of this influence. The latest of these is the National Science Foundation (NSF), until recently, perhaps, the least controversial of government agencies. Today, however, NSF is finding itself under attack by some members of Congress and some elements of the public for supporting certain types of research grants in the social sciences and the development and implementation of science education programs that some find objectionable on the grounds of the values they teach.

As many of you may know, this controversy over the support of grants deemed unworthy or even detrimental prompted one Congressman to introduce legislation that would require every research grant approved by the NSF to be reviewed and passed on by Congress. To the alarm and consternation of many, this amendment to the NSF authorization bill was approved by the House of Representatives. Fortunately, it was voted down in conference. However, I think it has left a permanent mark on the scientific community, because it has served as a warning as to the kind of attitudes and restraints it may face in the future.

This Congressional activity and other public reactions indicate that a good part of this new outlook springs not only from economic interests, or even from concerns with the risks of physical harm to the environment or people, but from a new combination of growing humanism and increased activism on the part of many segments of our population. This combination, forcing greater attention in the public area to ethical and human value considerations, is an outgrowth of the many movements and attitudes we saw develop in the 1960s and early 1970s. It had its origins in the youth movement, war activism, civil rights, environmental interests, and the new consumerism. It will have an increasing effect on the support and conduct of science, and I think most scientists are beginning to recognize this. As in many other cases of new influences, it will have its good and bad effects. Essentially, it is vital that science does serve the highest interest of society and does contribute to the fulfillment of human values. I believe that the science community for the most part is acting very responsibly and responsively in this direction. In many areas of research, such as genetic experimentation, atmospheric work, and the effects of chemicals on human health and the environment, it has taken the lead in initiating measures to place human concerns above all. In addition, ethical and human values in science and technology are themselves the subject of research programs being supported by the National Science Foundation in conjunction with the National Humanities Endowment.

But it should be realized that although there are certain values and ethical codes of a universal nature, there are also values that are more closely associated with the tastes, likes and dislikes, habits, and culturally induced beliefs of various individuals and groups attuned to certain so-called lifestyles. In a democratic society — and particularly one of growing advocacy and activism — there are bound to be many conflicts over these. Science and technology, with its increasing influence on life in general, will certainly be caught up in many of these. If this is the case, it may be essential for means to be worked out to establish some broad codes of conduct and values by which science and technology can operate to maximize human benefits within a framework of some type of consensus value scale. It seems to me that we must do this to avoid being paralyzed by a kind of case-by-case value judgment of all we do. This does not mean that technology assessments and risk/ benefit studies of individual concepts should not continue to be conducted, nor does it mean that science should not maintain a most profound sense of responsibility toward safeguarding society from possible errors on its part or misapplications of its work. It does mean, however, that we must find a way to avoid having a "tyranny of parochial interests" when it comes to the possibility of advancing the general good through scientific progress.

There is no doubt that this problem — the matter of balancing the human values of individuals with the technological and social imperatives of the larger community — is going to be one of the major problems for democratic societies around the world. There are, as I point out in a moment, tremendous pressures forcing this situation, and the world with its growing and dynamically related problems can ill afford to suffer "a failure of nerve" or "a failure of imagination," as Arthur Clarke terms them, in using science to help build a better future. But this is a possibility if conservative attitudes toward new knowledge and change come to dominate the scene.

Having discussed a number of problems and influences that may affect the future of science, many of which could act as restraints on its growth, let me turn now to the reasons why I believe science must, and will, grow. By that growth I do not mean simply a larger expenditure of money and manpower devoted to it, but the development of new directions and paradigms to advance it. I see three major groups of pressures that will force an increasing development and use of science and technology for man's survival and change. These three are related to the thinking of a parson, a politician, and a psychologist — Thomas Malthus, David Ricardo, and Abraham Maslow.

The first, and perhaps the dominating pressure, is the Malthusian. Here I refer to more than just the need to balance population and food supply, although that remains the most basic requirement. The Neo-Malthusian concept encompasses the production and consumption of all resources. No matter how we may disagree about man's ability to increase production, about the extent of the world's nonrenewable resources, about the Earth's carrying capacity environmentally or thermodynamically, or any such thing, we must agree that the global rate of population increase today does exert an enormous pressure on civilized man. It is a physical, social, and moral pressure. And no amount of intellectualizing about *triage* or a lifeboat ethic is going to reduce it. I do not see this nation or any other advanced country "writing off" any desperate people without a major effort at emergency aid. But even more important, I think we will see a greater attempt to help all struggling nations develop means to help themselves. And this is where science and technology will play a big role — in discovering, developing, and transferring the best means for peoples in different lands and facing different sets of conditions to work out their own problems and their own destiny. This is the approach increasingly being emphasized today. One can see it in the new aid and advice going out to the developing nations, in the new educational programs, and in the policies of the United Nations, the World Bank, and other international development organizations. It should prove far more successful than attempts to impose westernized forms of agricultural and industrial development on areas where they may be physically or culturally unsuited.

Malthusian-related pressure will continue to fuel the need for more science and technology, even beyond the point where the so-called "demographic transition" takes place (where the economic and emotional need for a large family declines) and the birthrate levels off. This can be seen in countries in which the birthrate is approaching or has already reached zero population growth. In these cases the labor force will continue to grow, as will the formation of new families. Thus certain changes will force new economic growth, even though the number of new mouths to feed eventually levels off. This is evident right here in the United States where, on the basis of children already born, the next decade will see an increase of 25% in the labor force and 34% in the number of new households.

In addition, this population age shift will be responsible for a 61% growth in the number of consumers in the 24 to 34 year age bracket (3). Such a shift might not have been very meaningful in a world like that of the Middle Ages, where people expected their rewards in the hereafter. But our global civilization today is increasingly temporal. Its instant communication and almost as rapid travel have sown the seeds for the kinds of pressure that will drive science and technology ever further. We might term that pressure Maslovian (after Abraham Maslow) because it forces men to seek fulfillment of a hierarchy of needs that go far beyond those of day-to-day survival.

This Maslovian pressure is related to the revolution of rising expectation that we have witnessed in the past few decades. There are several variations on this theme. In the United States and other advanced nations Daniel Bell calls it the "revolution of rising entitlements" because of his feeling that many people now no longer merely expect a better life but believe that such a life is literally their birthright in a potentially affluent world, and they demand it. In contrast to this, recent international economic and energy problems have prompted the former UN delegate Charles Yost to speak of the "rise of falling expectations" and warns that the disappointments ahead for those who are frustrated in their drive for a better life will provoke rage and new political violence and upheaval.

I do not want to predict how this will all come out, but I can assure you that one of the topics that will be foremost on men's minds in the years ahead will relate closely to the matter of a better distribution of the world's wealth. The effect that this is going to have on science and technology will be significant. It will cause, as I mentioned before, a focusing on methods of agriculture and industry suited to the developing regions. Among many other things, this will stimulate all the multi-faceted research related to the cultivation of arid and tropical lands. It will increase the focus on health and nutrition in these areas. It will force a more intense investigation of the natural resources of these lands. In addition to the finding of better ways to use these as commodities, it will emphasize the innovation of labor-intensive methods of manufacturing them into products for which there will be world markets.

All this is going to pose a great challenge to science and technology, to both the advanced and developing nations who must learn to work together ever more cooperatively. I see no alternative to this other than an increasingly explosive and catastrophic situation.

Related to the Malthusian and Maslovian pressures on science and technology will be a pressure tied to a theory of David Ricardo. This essentially is the idea that as demand on resources grows and they become scarcer, less accessible and lower in quality, the capital costs of finding and developing them rises, until eventually such costs are prohibitive. Then a major decline in the standard of living sets in. This is a major factor in the Meadows' model in the Club of Rome project "The Limits to Growth," in which Meadows sees the world running out of capital to develop resources long before those resources are actually physically depleted. He discounts the fact that science and technology have long been engaged in a running battle with Ricardian economics and have managed to keep well ahead of the game. For example, we have managed to find and use lower grade ores at lower costs. We have developed cheaper substitute materials to replace costly resources. We have improved our efficiency and productivity in ways that have reduced material, labor, and capital costs. We have shifted to new technologies and social systems that change our relationship to resources entirely. The question now is, at what rate can we continue to do this? Meadows and his followers argue that because the doubling time of our population growth and resource use is now so short, we cannot avoid a catastrophic decline within the next 50 to 100 years without establishing a much slower rate of growth and a move toward stability.

Whether or not you agree fully, partially, or not at all with the Meadows' models or their conclusions, it seems to me that science and technology are now in a race against time. It is a race that must see us develop, at a relatively higher efficiency of resource use and a lower energy level, a standard of living high enough and well distributed enough to achieve some social and political stability throughout the world for the next 25 to 50 years. In achieving this, we will be enhancing the possibility that world population may level off between 7 to 10 billion people. During this period, we in the more advanced and affluent nations will have to sacrifice enough of our standard of living to support intensive research and development into a variety of innovative — perhaps radically new — ways to operate the world. These will include entirely new energy technologies, agricultural and industrial systems, and national and global political and economic arrangements. The truth is becoming clearer each day that although we may not have reached the limits to growth yet, we have just about reached the limits of conducting our lives in the same old way. From this standpoint, Mesarovic and Pestel are correct in the Second Report to the Club of Rome: Mankind *is* at the Turning Point. We must restructure and redirect our attitudes and efforts in the light of the organically interdependent world that exists today. To do otherwise, to try to continue along the same line of the past quarter century, is to court catastrophe.

Before shifting to some thoughts as to the directions science and technology might take at this turning point, let me state a few basic social imperatives that must underlie this movement. I believe it is essential that we pursue and achieve three broad goals:

First, that we maintain peace throughout the world and intensify our efforts at arms control and limitation, and eventually disarmament. The world can no longer (if it ever could) afford "guns and butter." A new meaning must be given to security tied to global development.

Second, that we establish a much higher level of political and economic cooperation among the nations of the world. A global economy exists and must be treated as such.

Third, that we further increase our international cooperation in science and technology — in the kind of joint research that will allow us to find mutually acceptable solutions to global environmental problems and in programs of technology transfer that will benefit all.

I will not dwell on these three imperatives, as I believe this audience is highly knowledgeable about them and well-tuned to their need. Instead, let me conclude with some brief comments on the directions science and technology may take in the years ahead. Most of these directions will be centered on two broad goals: (1) more fully establishing the boundaries — physical, environmental, and social — in which we can operate and (2) providing the knowledge capital that will allow us to operate within them. It is that knowledge capital, on which we have drawn so heavily in the recent past and which we must replenish with new ideas, that will allow us to compensate for declining physical capital and higher cost resources.

For example, let us turn to the field of food and agriculture so prevalent on the world's mind today. In many ways the production of food has become more energy and capital intensive. Modern high-yield agriculture depends to a great extent on chemical fertilizers and pesticides, on irrigation, and on mechanized production and harvesting methods. To bring more of the world's arable and potentially arable land into production will require considerably more of these resources.

This is particularly true as we try to develop the semiarid and tropical regions that present special problems. But though we will do this, our biological and biochemical sciences hold the key to many other ways to increase our food supply. For instance, they may help us convert a large portion of our agricultural products and other natural products not now edible to humans into a huge new source of food. New research in microbiology and enzyme chemistry is indicating that the more than 150 billion tons of cellulose waste produced annually with the world's agricultural output can be converted to food, fuel, and other valuable products. With the help of certain fungi and enzymes it is possible to turn this material into sugars, alcohols, amino acids, and a variety of other materials — organic chemicals, solvents, drugs, and antibiotics. In fact, an entire bioengineering factory has been visualized which, operating on very low energy levels and with minimum environmental impact, would use farm waste and organic refuse to produce all these and perhaps many other products and articles, ranging from animal feed and human food to polymers that could provide everything from building materials to golf balls (4).

The biological and chemical sciences are going to help us in our food situation in other ways. Research in nitrogen fixation could lead to the creation of crops that would reduce our heavy demand for synthetic fertilizers, thus reducing the chemical and energy resources going into them and their environmental impact, such as in nitrate runoff. Biologically integrated pest control, in which there is much interest and research taking place, would also relieve us of a growing problem by helping reduce the huge losses — more than 50% in some areas of the world — of food supply.

Through new research in plant genetics, soil science, hydrology, ecology, and many other fields, we must learn to support new types of agricultural systems in parts of the world that have resisted previous efforts at such developments. Where this may prove too difficult or costly, perhaps chemistry can help us make use of some of the existing vegetation as food.

One example of this has been the successful extraction of leaf protein which has been used on an experimental basis to relieve the protein deficiency of a group of children in India. More broadly speaking, I think we are going to witness a new outlook on the world's vegetation and plant life. Of the some 400,000 million tons of vegetation produced annually by the process of photosynthesis, man's yearly harvest of food and fiber amounts to a fraction of 1% (5). Of the some 250,000 plants known to man, we use fewer than 100 on any large scale for food (6). Thus there is a great deal more of nature's bounty we can draw upon if we learn to understand what she has to offer and how we can use it best. This is one of a great many areas in the life sciences where we must build more knowledge capital.

If we turn more to the study of biological systems in the future, I think we also will be studying and learning more that is very useful in the large systems in the atmospheric and earth sciences. We have witnessed the beginning of this in recent years in the initial efforts of the Global Atmospheric Research Program (GARP) in the tropical Atlantic last year. Over the next decade or so this big international program, conducted with the cooperation and scientific resources of so many nations, should give us a much better understanding of the interaction of ocean and atmospheric forces. This in turn will increase our knowledge of the generation of global weather, and together with other research in the atmospheric sciences could play an important role in our weather and climate forecasting. We realize now how necessary this is not only to the future of world agriculture but to many other aspects of our lives. In addition to the reduction in human suffering that more knowledge in these fields could effect, there are large-scale economic benefits that could be gained by a more precise understanding of weather phenomena and perhaps someday a degree of success in weather modification.

In a similar vein, much is to be gained in the earth sciences. Work in the relatively new field of plate tectonics and other geologic disciplines could reveal much information that would help us in the discovery of new reserves of mineral resources, as well as give us better control over our dealings with earth-quakes. Here again there is vital economic and human motivation to acquire and use this knowledge capital.

One area of scientific investigation to which we are going to be forced to devote more attention is that of hydrology. In recent years attention has focused on water pollution, and some success is being achieved by government and industry to clean up and protect our rivers, lakes, and estuaries. But for some time now, the world's water supply problems have been build-ing. We have been drawing increasingly on the supply of fresh water. Worldwide irrigation has grown tremendously. The de-mands of industry and cities in many areas are rapidly ap-proaching the limits of the supply of runoff water. In some parts of the world underground water supply is being dangerously drawn down, and in many countries there are growing satura-tion and salinity problems. In view of all these problems, the world water situation could reach crisis proportions even more pressing than the energy situation if not attacked by a concerted international effort. Recent water conferences have been sound-ing the alarm on this, and I hope it is heeded in time and with adequate cooperation and resolve.

In many of my previous speeches, I have stressed the belief that we could solve a good portion of our water, energy, and materials resources problems by greater efforts toward recy-cling. And I have spoken of these efforts as a movement toward a recycle society. Evidence is building that this idea is taking hold in many parts of the world. Many large-scale resource recovery facilities are already in operation or well into the planning stages in this country on a municipal and even a statewide basis. The reclamation business in the United States now exceeds $8 billion a year (7) and is growing constantly as new economic means are conceived to collect, recycle, and reuse materials that previously went to waste in landfill or ocean dumping operations. Methods are being developed not only to recycle metals more completely and economically, but to reuse synthetic rubber, plastics, and a variety of organic materials. This is true from the use of sewage sludge here as fuel and fertilizer to the use of cigarette butts in China, wherein the nicotine is extracted and processed into a product that treats some 1.5 million acres of farmland (8).

Science and technology in the years ahead should play a major role in changing and improving man's use of his resources. The era of abundance in terms of waste is over. Scarcities, the threat of scarcities, and even the contemplation of scarcities is going to change our ways of operating. This is particularly true concerning energy, a subject with which I would like to conclude. The energy situation is unique. It is the first time in history that the entire civilized world is affected by the threat of a diminishing resource. Regardless of the fact that some people presently see an "oil glut" in parts of the world or that large reserves of coal are still in the ground, we should expect the decline of the fossil fuel age within the next 50 to 100 years.

In view of this we must act now, for a number of reasons. The first lies in the recognition of how essential energy is to every aspect of life. Previously the availability of energy resources and its relatively low cost made us forget this. In addition to this threat of reduced availability of energy and its rising cost, there are other reasons that will drive us to use our science and technology to develop alternative energy resources. One is the environmental impact associated with the further exploitation of fossil fuels. Another is the fact that these hydrocarbons are an invaluable source of chemicals for industry. As I once pointed out, no less an authority on oil than the Shah of Iran has reminded us that there are some 70,000 derivatives from petroleum today (9). Still another reason is that it takes a long time — not years but decades — to properly research, develop, and bring into full commercial use new energy technologies. They cannot be introduced full-blown overnight, no matter how good they look on paper or how well they work in the laboratory. Also, it takes a long time to make all the social and economic adjustments in a large-scale energy transition.

These and many other facts would indicate that over the coming decades, we must begin to change our relationship to our energy resources and technologies. To make the required transitions we must carry out, particularly in the United States, a concerted effort in energy conservation. This can be done, and if carried out intelligently, without any great reduction in living standard. A reduction of energy waste in industry, in transportation, in our homes and office buildings could help to cut our energy growth rate from its more than 4% to between 2 to 3% with little economic hardship.

But some initial economic sacrifices will have to be made to develop our energy alternatives — and this is true for the entire range of them, whether they are nuclear, solar, fusion, or any combination of these. The current impasse in developing legislation to set long-range energy policy for this country lies principally in the fact that the political courage is lacking to tell the public it will have to pay in a number of ways to assure its energy future. Until we get by this impasse, our progress will be slow. Once past it, we may see remarkable advances.

Associated with this is a final thought with which I would like to conclude: that the energy problem epitomizes the great dilemmas we face in using science and technology to advance the quality of life for the human race. Our success in science over the past few decades has, in a sense, fostered many new problems for the world. But it also has given many of us a false sense of security, an idea that science moves us toward a utopian, problemless, riskless society. Nothing could be further from the truth. First, because man, not science and technology as such, is always accountable for the choices that either enrich or diminish the quality of life. Second, we live and always will live in a *dynamic* situation, amid problems whose solutions will breed other kinds of problems, and in a society where the leaps of progress will be proportionate to the risks taken. Even within the bounds of a "steady-state society," a "no-growth society," or any other scheme of population-resource-energy equilibrium we might achieve, there always will be change and creative growth that will challenge the human intellect. There always will be dangers, risks, and increasing responsibilities that will drive us toward a new level of excellence in all we do or try to achieve. This is the process of human evolution at work — a process that started with man's ascendancy and will continue for some time, how far and to what end I will leave for you and others to speculate upon.

But we must believe that it will all turn out well, and work constantly toward that end. As a scientist, I believe as Edwin H. Land once put it for scientists and engineers: "optimism is a moral duty."

REFERENCES

1. Louis Pasteur, Inaugural lecture at Douai, University of Lille, 1854, as excerpted in *A Treasury of Great Speeches*, Edited by Houston Peterson, Simon and Schuster, 1954.

2. "Prices for physics equipment outstrip inflation," *Physics Today*, Jan. 1974.

3. Remarks by Thomas O. Paine, Senior Vice President of General Electric Co., Fairfield, Conn., at the American Power Conference, April 22, 1975.

4. Concept attributed to Dr. T. Meloy, Director, Division of Engineering, in National Science Foundation, Engineering Program Review, Jan. 1975.

5. R. Katz, "The Politics of Doomsday," *CERES, FAO Review on Development*, Jan.–Feb. 1974, p. 28.

6. N. Pirie, *Food Resources*, Penguin Books, 1971, p. 59.

7. *Environmental Science and Technology*, August 1972, p. 700.

8. *New York Times*, January 26, 1975, p. 61.

9. *Time Magazine*, April 1, 1974, p. 41.

DISCUSSION

General Comments

Dr. Libby. Dr. Seaborg pointed out that certain factions of society are very pessimistic about mankind's future and that he was an optimist, and that he thought most scientists would be in the optimist class. I find myself very definitely in agreement with Dr. Seaborg on that point. It's very difficult, as he says, to imagine doing science successfully without an optimistic approach, since all of us who do science have some hope or thought that our work will prove useful and beneficial to mankind. The doing of science is different from the doing of technology, which is the step between science and application. But the doing of science is motivated, at least in my case, by the desire and hope that the results will be useful. So I think Dr. Seaborg was entirely correct in saying that science is, per se, optimistic.

Dr. Harvey. Well, I found that aspect of Dr. Seaborg's talk the most problematical. I think we have every reason to call for hope, but I think one has to ask what the grounds for optimism are. One can hope for mankind, but the history of mankind doesn't lead me to be optimistic.

Two things, for example: First, Dr. Seaborg made the point that everyone is going to have to sacrifice his own personal expectations at just the time at which rising expectations all over the world are the primary and fundamental attitudes of us

---

Editor's Note. Asterisks denote a break in sequence or a change from one discussion panel to another.

all. I see no indication whatsoever that the American people, or, indeed, people over the earth, are going to sacrifice or moderate their expectations. And I think this is a very, very important thing we have to keep in mind.

Second, I see no indication that the loyalties of the American scientific community extend beyond the global limits of the United States. Indeed, it seems to me to be that American science, on the whole as a profession, has put itself in the service of national aims without really raising fundamental questions as scientists about those aims.

Dr. Cournand.   I have been a physician, applying physiology to the study of disease. A physician must be optimistic, not only in his profession, but because he has examined the problem of life and found that in order to live one has to be optimistic. I think that in every situation of man in this world one has to be optimistic in order to survive.

I have another reason to be optimistic and that is because scientists *have* become aware of the importance of social problems and of their responsibility in solving social problems. I do believe that some of the difficulties, which we may avoid in the future, occurred originally because many scientists were completely unaware of the effect that science has on society. Scientists are not only becoming aware of social problems, but they are becoming aware of economic problems and they realize that these problems can be solved only in a global fashion. That is, they have developed, as was expressed so well this morning by Dr. Seaborg, a terrestrial attitude.

Dr. Weller.   I represent the small subset of physicians that deals professionally with the health problems of the developing areas of the world. If we weren't optimistic about the future, there would be no point in practicing our profession or having any interest. Certainly, the world has tremendous problems. But man is at a turning point and we must, if we are to continue to progress, rethink what our priorities are and realize that we all are going to have to make sacrifices. But things are not without plenty of reason for optimism.

When I started teaching tropical public health at Harvard in the early 1940s I talked about a life expectancy in the developing areas of the world of around 37 years. Now mean life expectancy in the developing areas of the world is on the order of 50 years or better. We have created some problems, though. Deaths and births are out of phase. When I started teaching there were two billion human beings on this globe. Now there are about four billion. Dr. Seaborg mentioned that as we approach zero population growth in the United States, we're going to have more and more people coming into the working age and we're going to need more jobs. The converse is true in the developing areas of the world, where the world is rapidly growing younger. We're soon going to have more than half the world's population under 18 years of age. We're going to have to accept this imbalance for a while, but one can look for signs of optimism.

In Costa Rica, for example, there's a relatively stable government. There's almost universal literacy, and if one looks at the population pyramid, you'll note that in Costa Rica the bottom tier, the 0-to-5 age group population, is now smaller than that just above it. In other words, things are beginning to improve in Costa Rica. As there is social and economic development, there are fewer children being brought into the world. This is the job of the scientist, to expedite, to work on the health problems in the developing areas of the world, so that we can get rid of premature death. If this is phased in with social and economic development, then the scientific effort will be productive. But science by itself, without better education, without better agriculture, without better industry, without the whole complex of social development, won't succeed — we're working in partnership.

* * *

Dr. Walton.   I find very little in Dr. Seaborg's address with which to disagree, but I would like to emphasize one of the topics that he mentioned. This is the question of our irreplaceable resources. We have been using up these irreplaceable resources, and these are capital resources. We've been living on capital. We have been a spendthrift generation, and everybody knows what happens ultimately to the person who lives on capital and not on income — at some point you come to the end of your capital.

Dr. Seaborg has suggested that as far as fuels are concerned the end will come in 50 or 100 years time. I believe that these times of comparative plenty ought to be used in order to gain the knowledge which will enable the human race to survive when times of famine come later on. I believe, and I think a great many people will agree with me, that the human race is the most important thing that we know of in the universe today, and that it's of the utmost importance that this race should persist. In this time of plenty we ought to devote our energies to research in order to fulfill our destiny in the years to come. I would say that I regard it as a moral, perhaps religious, duty to devote a considerable part of our resources to acquiring the knowledge which will be necessary for the survival of man.

*   *   *

Dr. Barbour.   One ethical issue that Dr. Seaborg raised which seems to present more serious problems to me than to him is the gap between the rich and the poor countries. Technology tends to increase this gap. I see this as an ethical problem, in that the need for social justice will become more acute, particularly if one sees a need for a reduction in growth on a global scale. If growth slows down, then the question of inequality will become increasingly serious, since the possibility of nuclear blackmail will increase the certainty of world chaos and threaten the future stability of the world. As an ethical issue facing affluent nations it seems to me that an even more radical reorientation of priorities is required to effect a solution.

Dr. Seaborg. I don't find myself disagreeing. I tried to make the point that we should increase our efforts in connection with the developing countries, not only in energy but in transfer of technology. We will have to make sacrifices, and accept some reduction in the standard of our living. I used the word "catastrophe" to describe the alternative to this.

* * *

Dr. Kuznets. Dr. Seaborg's presentation dealt with science and technology in a very broad range, extending from the pure levels of basic science to what might be called rather specific technological problems and tasks. Because of the coverage of such a broad range, a number of questions arise as to the feasibility of the kind of hopes that Professor Seaborg expresses and as to the warrant for the kind of optimism he entertains.

The pressures of which Professor Seaborg spoke are rather unequal, as they are viewed with reference to these different elements in the range from pure science, theoretical, to technology, specific. The question as to what determines the use by society of resources at different levels of this range has not been answered yet. How does society determine how much to spend on technology having to do with new energy?

Now one reason for my not being able to share Professor Seaborg's optimism is that technology is essentially determined by the society within which the technology is originated. It is directed at the problems of that society, not necessarily the problems of the world. There is a real question as to whether technologists in the United States are either qualified or will be financed by our society to work on problems that our society does not view as its immediate concern. There are two big questions that have to be faced. One has to do with transferability of scientific and technological resources for what may be called national problems to international problems, and the second has to do with the political feasibility of expecting that society will agree to the kind of massive transfer of resources that would be involved in a real attack on the problems, technological and others, peculiar to the poorer parts of the world.

Dr. Schwinger. While Dr. Seaborg spoke very eloquently of the importance of the development of science and technology for the solution and amelioration of the world's troubles, I think he neglected to give sufficient attention to another, and to my mind perhaps even more basic, reason for developing science that I might call, in the present context, its contribution to man's spiritual values. By this I mean the role of science in developing man's understanding, his comprehension, of nature. This, after all, is what fundamentally drives people to do science. Nobody becomes a scientist because he wants to help settle the difficulties of the world in meeting its energy problems or anything of the kind, but because he wants to know — to understand. And science needs basically to be supported for that reason.

Too often scientists, in justifying what they do and why society as a whole should support that effort, point to all the practical benefits that come from it. It's well known that for every dollar invested, ten dollars — or whatever the number may be — will return to the community as a whole. But that is not the fundamental reason that science is done. I think science should be supported basically for the same reason that symphony orchestras are supported. There is no economic return from this, but we all know its cultural importance. So, I would emphasize this other side of the coin — the spiritual, the cultural values.

* * *

Energy

Question. Yesterday in his airport interview Dr. Brattain suggested that nuclear energy should only be used as a stop-gap measure. What do you see as the future of nuclear energy and what of the future of mankind if nuclear energy proliferates?

Dr. Seaborg.   I think it's difficult to say whether I disagree with Dr. Brattain when he says a "stop-gap measure." I would certainly replace nuclear energy with solar energy or fusion after it has fulfilled its role, but that will be some time in the future, sometime in the twenty-first century. I think nuclear energy is a form of energy that we are going to need, when you analyze all the actually available forms of energy. It will fulfill a role for a number of years, and I think it would be no surprise to this group or audience if I indicated that I thought it can be used safely and successfully. I might say that Dr. Hans Bethe has in recent years given this as much thought or even more than I have and he might want to comment.

Dr. Bethe.   I would indeed like to comment. And I'd like to comment in three ways. First of all, I agree entirely with Dr. Seaborg that nuclear energy is necessary. In our present situation we have only two sources of energy that we can fall back on in the face of decreasing oil and natural gas, and these two sources of energy are coal and nuclear energy. Coal, obviously, has great problems because you need so much of it, and just to get it out of the ground and to the place where you want to use it is difficult.

Secondly, we do not know whether fusion energy is possible at all. Experiments are now going on which possibly within ten years will answer this question, and I think you have to visualize what this means. Nobody can tell at this point whether you can get useful energy from fusion on earth. Suppose the answer is positive, then there is the question of economics. I have looked at the University of Wisconsin, where they are doing some preliminary engineering, and they tell me it will cost at least 50 percent more than present nuclear reactors. If people say 50 percent more when they work enthusiastically on the project, I think you have to be prepared that it may be twice or three times as much as the present nuclear power. Therefore, it is quite possible that fusion will be feasible, but yet economically totally unacceptable.

Third, while solar energy is clearly around us and while it may be useful for the heating of houses, I am very, very doubtful as to whether solar energy for the generation of large amounts of energy will be economical within 100 years. So I would say it slightly different from Dr. Seaborg. I would like to say that nuclear energy has to be used until something better is proved both feasible and economically competitive. And whether this takes 25 years or 50 years or 100 years, I think none of us can tell. It may easily take 100 years. I just don't know.

Dr. Seaborg.   You know, I stated it very carefully. I said sometime in the twenty-first century. That gives me 125 years.

Dr. Barbour.   Certainly the figure of 50% that the Wisconsin group uses isn't prohibitive. It seems to me that it would be a small price to pay if one is choosing between the nuclear route and the solar route in the long run, particularly if one takes nuclear proliferation into account. After all, nuclear technology has developed to the point it has largely because it received tremendous financial support; first, because of its connections with weapons, and then because it received a lot of development money. Nothing even remotely comparable has been spent on solar technology. One might then expect a considerable reduction in cost if one put anything like that support into it, and I would wonder if that isn't a small price to pay when compared with the dangers in the nuclear route.

Dr. Bethe.   I am unhappy to monopolize the conversation, but

I would just like to comment on the point which you made, that if you spend enough money, then you will surely find a way to solve the solar energy problem. I think this goes right back to one of the first remarks of Dr. Seaborg's, namely, that the public has come to expect miracles. The public has come to expect that when the scientific-technical community is given enough money and enough manpower, then any problem can be solved which is posed to the community. This isn't at all how it works and it isn't at all how it has worked.

` When, for instance, radar was developed, it was developed because some people had a good idea which was founded on previous basic knowledge of electrodynamics. Somebody had the good idea that you could use electromagnetic waves to see flying objects in space. Once you have a good idea and then have manpower and money, then you can develop something. But in the field of solar energy, as used for large generation of power, there has not yet been a good idea and I don't see where it will come from. People have had at least a dozen ideas, and the best of them are being supported with a lot of money, as much money as makes sense in the early phases of research and development. But the best idea isn't good enough. This best idea which I talked about will give a cost investment not one and a half times, not three times, but five times the cost of nuclear power at the least.

Dr. Seaborg.   Let me just break in here, and then I want you to continue. To illustrate that it isn't only the cost, the methods that are being suggested now for producing electricity from solar energy use more energy in producing the component parts than you get out, and that presents a basic problem. You have to get around that. It isn't only a matter of cost.

Dr. Shockley.   Glenn, a clarification on that. You continue to get the power out for a long time. Do you mean it would cost more to make the facility than you would get out of the facility?

Dr. Seaborg.   Yes, amortizing it over a reasonable lifetime.

Dr. Bethe. Most of these components will deteriorate. Take photoelectric cells. They certainly deteriorate quite fast. Take mirrors, which are in the solar facility I like best. Mirrors get dusty and have to be maintained, but after a while they may break and they have to be replaced. They certainly don't have a very long life. So, I think I agree entirely with what Glenn said.

Dr. Lipmann. As a biologist, I should like to know how much energy can be made by solar energy in plants. There are enormous amounts of energy stored in trees, either those presently growing or fossilized ones. They are used today as wood, coal, and oil for energy production. On the other hand, energy deposits are still being made all the time photosynthetically in plants. I think it is very encouraging to see that some technologists have become interested in biophotosynthesis and in biological process principles in general. I would like to point out that the brain is considered by many to be a computer, and indeed, the most compact and effective computer that exists. Therefore, the utilization of biological process principles, particularly the dominant use of wet processes which are much easier to miniaturize than the hardware of our technology, may give new routes for providing energy production and for other branches of technology.

Dr. Seaborg. I think that agrees with what Hans said, and I would put it another way. I think as far as solar energy is concerned we haven't discovered the method we're going to use yet. This doesn't mean we can't find it. It points to the need for more research. Maybe some kind of a catalyst will cause water to be split by sunlight, giving off hydrogen, so you could go to a hydrogen economy. I don't know what the method will be. I, at least, tend to believe that we haven't discovered the method we're going to use. And this illustrates the need for more basic research.

*   *   *

Values and Science

Dr. Bloch.   There is one point in Dr. Seaborg's talk which I
would like to comment upon. I quite agree with him in many
ways when he describes the pressures and constraints that are
undoubtedly going to be exerted on science and technology in
the future. But there is one point to which I want to refer. He
talks about the values and ethical codes which will be posed.
He speaks about the fact that there are habits and culturally
induced beliefs held by various types of individuals, and says
that undoubtedly from that will arise certain conflicts. Then he
says, "If this is the case it may be essential that means are
worked out to establish some broad codes of conduct and values by which science and technology can operate to maximize
human benefits within a framework of some type of consensus
value scale."

Now this sounds fine, but I believe that at this point one will
have to distinguish between pure science and technology,
which I consider the equivalent of applied science. I believe
what Dr. Seaborg says about the code of values is probably
possible and quite possibly necessary in regard to technology;
but I totally fail to see how such a code, no matter how broad,
could be established in the realm of pure science. Let me give
you an example. When Einstein developed his theory of relativity in 1905, out of that came the very famous equation that
you all have seen in the newspapers, $E = mc^2$, which means the
equivalency of mass and energy and implies, in a certain sense,
that mass could be transformed into energy. Well, about 40
years later this became a reality. It became a reality through the
atomic and nuclear reactions and eventually found its most
spectacular manifestation, unfortunately, in the nuclear bomb.

Now, I ask, *should* there have been a code of ethics in 1905
which would have told Einstein not to develop his theory of
relativity so as to prevent the development of atom bombs? You
see the absurdity of it. It is quite clearly impossible. I believe
also that it is not the responsibility of the scientist, who works

at the level of the research that Einstein did, a man who investi-
gates the intrinsic structure of nature; it is neither his responsibility nor is he capable, nor is anybody capable of telling him
what is good or bad, assuming people can agree on what is
good and bad. So, therefore, I would like to say that I accept Dr.
Seaborg's postulate only in a very limited sense.

Dr. Dean.   I'm concerned with Dr. Bloch's opening statement
that pure science should be free from, apparently, any moral
constraints, and that simply the pursuit of knowledge should
be the only criterion. I certainly sympathize with that, but what
if you cited the Manhattan Project as an example instead of
Einstein? It seems easier to criticize the science that went on in
the Manhattan Project. I wonder if Dr. Bloch is content to say
that utterly no moral restraint should be applied to pure science
research — that the scientist should be willing to hand over to
technologists, for whom moral criteria are apparently applicable, *any* information that is generated by pure science.

Dr. Bloch.   Yes, I think I shall stick to my guns. When you
mentioned the Manhattan Project this does not fall under the
category of pure science. This is a development application of
some phenomena which were discovered before without any
ideas whatsoever with respect to atom bombs. In fact, if I may, I
will even say something in defense of the Manhattan Project,
because it seems as if we were quite clear — this is good and
this is bad. I myself worked on the Manhattan Project and the
reason was very simple. We had very good reasons to believe
that Germany, under Hitler, would develop an atom bomb. In
that case, not to prepare a counteraction at that point just
seemed too dangerous. Fortunately, we were wrong, but we
did not know it. Germany did not succeed with that action, and
now, of course, we have the stigma of having developed this
weapon to annihilate mankind. But I must say that most of us
entered that project not with the intent of evil, but to prevent
evil from happening.

I don't quite know what Dr. Dean means by moral obligation in pure science. Pure science is neither good nor bad. It is motivated by exactly what Dr. Schwinger said — it is motivated by an understanding of nature. This, I think, is a defensible human drive. It is not only defensible, but as Dr. Schwinger said, it is one of the great contributors to human civilization and human culture. Shall one say that because of the possibility of abuse, people should stop thinking? It is impossible, and, therefore, I say to Dr. Dean, yes, indeed, I do not see how you can impose what you, Dr. Dean, or somebody else may call ethical constraints upon pure science.

*   *   *

Question.   Considering the risks involved in future scientific discoveries, does science have within its own nature the ability to decide what avenues it will pursue, or does it need the direction of society at large?

Dr. Cooper.   The question of social directives for science is, I think, one of the most important that we face because of the social and political pressures now placed on us. I consider these pressures one of the greatest dangers to science as we understand it. Let me explain why. We are faced with large social problems that have to be solved, and we should direct some of our efforts to solve these. I'm not opposed to that. But there is an aspect of science in which there's a slow development of the store of ideas — new ideas, new insights. That kind of science is not done by external directives. It's not done by social pressures which tell you there are certain problems that have to be solved.

I would like to explain in more detail why that is. If there is a problem that has to be solved, for example, if one wants to find alternative sources of energy, one might take from the ideas that already exist and build something. That's what I call technology. This has been the nature of the huge programs with which we're familiar, such as going to the moon and the Manhattan Project. They can be done with men, money, materials, and the ideas that exist.

It is in the production of new ideas where it is very difficult to give that kind of directive, because often the most important new ideas for one field come from an investigator in an entirely different field. The decision as to what one would work on comes from an internal directive, that is to say, the scientist himself perceives what is an important problem for the field. This is very similar to an artist or a musician deciding what will be an interesting painting or what will be an interesting piece of music. It's very hard to define; the person who is doing it has a sense of what the important problem is. Now if you crush that, and it is now being crushed to a certain extent, what will happen is that we will deplete this source of new ideas — the real source of breakthroughs.

* * *

Question. How is the ethical, humanistic viewpoint of science as opposed to "science for science's sake" going to affect the science of the future?

Dr. Huggins. I claim there is no ethical, humanistic view of science. Science *is* science for science's sake. I'm a purist in this field, and one pursues it in that line and not for the uplift of mankind.

Dr. Yang. I would like to disagree with my distinguished colleague. I would think that in a scientist's day-to-day life, when he is in the laboratory, when he is completely absorbed by the problem at hand, it is difficult to envisage that he constantly keeps in mind the larger context of his research. He is worried about how to substitute one atom for another one in a specific reaction; he is worried about one term in his equation where it should be a plus sign, but it comes out always in the negative sign; he is worried about these very mundane perplexing problems.

But I think that all of us, whether a scientist or not, in all human activities, in fact, are dictated in some sense by a certain frame of mind. I cannot believe that that frame of mind does not involve doing something for society. On the operational level, in the day-to-day contact, one may not be focusing on the social aspect of one's scientific endeavors. I do not believe, however, looking at the larger context, that this social responsibility of a scientist, the social significance of what he does, does not have a determining influence on the work of a scientist.

Dr. Huggins. If I could respond to my distinguished friend. What one does determines philosophy. I work with phenomena. I work in cancer research. In cancer research you must shut the door to Aunt Nellie's problem. Our work is strictly confined to working out a crossword puzzle made by oneself. One sets up a problem, then solves it. I hope and pray that it will help Aunt Nellie, but whether it does or not doesn't enter the question. From our standpoint there are no humanistic values in day-to-day operations. Maybe over the long run there are.

Dr. Yang. But that's precisely what I was saying. There is a value judgment in every field. That value judgment cannot be divorced from the value judgment of society. When you choose to study cancer, in your day-to-day operation you, of course, try to attack a specific problem. But the fact that society supports cancer research means society has a value judgment on the longer-range perspective of your work, and I cannot believe that that does not influence your work. If we sit back and ask, "What is the long-range significance of the problems we tackle?" or "What is the long range value given to the result of each piece of research work?" I believe that the fundamental and final value judgment does not rest with science for science's sake, but rather whether science is useful for mankind.

Dr. Kusch. I am restive with the word "useful." It seems to me that there are human activities which mark one's humanity. I submit that knowledge is intrinsically of value but in a different sense than the potential cure of cancer or the promise of perpetual motion, or something of this sort. It has an intrinsic value — it is intrinsically humanistic in the sense that it exercises the best qualities of man to discover the world, in the present context, that God gave us. I think that it has a value quite outside of its social context of providing more energy or doing something or other.

Dr. Mulliken. I agree with Dr. Kusch that scientific knowledge should be regarded as a cultural value. Of course, that's something human. I don't know what could be meant by science for science's sake. Science for science's sake means science for human knowledge. It's all human. We somehow can't get away from that.

Dr. Cobb. We get to the very old question of the good and the true and the beautiful. We haven't talked about the beautiful and we'll drop that for the moment. It does seem clear to me that science is a pursuit of truth, and nothing could be more humanistic and more intrinsically valuable than the pursuit of truth. I want to support that very strongly.

On the other hand, speaking from the side of the traditions of philosophy and theology, there is an ancient struggle between the concern for truth and the concern for the good, as defined in some way. And I fear that if scientists as human beings would themselves take the strictly purist position, then society will decide that it has to decide what the scientists are going to do. In other words, if a scientist says it's indifferent to him whether he's seeking the way to destroy humanity or to cure disease, somebody else will step in and say there are some truths more important to pursue than others. I hope that the scientific community will not itself abdicate participation in the decisions about the kinds of scientific research that are more appropriate to push forward than others.

Dr. Yang.   I find myself almost in the position of a minority of one. But I think part of this is due to a confusion of language. In the longer perspective of the history of science, society and scientists together have formed a sort of value judgment. I submit that this value judgment is deeply embedded in the interaction of science with society. There are truths and truths, but what truth is more important is determined by society.

Dr. de Duve.   I think it's a false problem to consider, as an ethical problem, the question of whether we should or should not pursue truths or try to satisfy our curiosity, because it would be the same thing as saying that it is an ethical problem to decide whether we should or should not breathe. I think it is part of our nature as humans that we are curious and we want to know. And I think it's one of our major attributes as humans that we want to know more about ourselves and about the world around ourselves, and nothing is going to stem that curiosity. So, I don't think that's an ethical problem.

It has been suggested that ethical problems may arise in the choice of subjects to work on. There again I think it's a false problem, because when you choose to work on a given subject it's because you want to know about that particular aspect of nature and not because of some possible application. Since you do not know what you're going to discover, you are obviously not in a position to decide what will be the good or the bad application that will stem out of your discovery. So, I don't think the choice of topics is an ethical problem, I think the ethical problems arise in the choice of methods. I work in the life sciences, and as you know there are a number of experiments that one simply should not do. There are experimental approaches that we can consider unethical and which should not be used to arrive at a solution to a problem, even though they may be the easiest or the best means to arrive at a solution.

The other ethical decisions arise when it comes to apply the knowledge. Here, obviously, is a problem that we share with the rest of mankind. The scientists do not themselves decide on the applications, but they may decide to develop applications, and this presents an ethical problem for the scientist. I think this is particularly true in the field of physics and in the field of the life sciences, where you must decide whether you should help in developing applications of knowledge.

* * *

A Personal View
of Science and the Future

by

POLYKARP KUSCH

In talking about the future of science, I am something of
a purist. Science and technology are virtually synonymous in
the public mind, and in fact are often nearly inseparable from
one another. My emphasis, however, is on science as the discipline that seeks understanding of the natural world. In the
popular mind there is a pervasive misunderstanding of science.
The purpose of science is believed to be the creation of new
technologies. If new technologies appear to improve man's lot,
science is in high public esteem; if, in the long pull,
technologies have damaging consequences, science is de-
nounced. It seems appropriate to begin with a comment about
technology.

The difficulty in offering brief remarks about the future of
technology is that technology is a much more complex enterprise than science. It is formed not only by science but also by

economics, sociology, and politics. Technology as a functioning reality, as distinguished from an idealized statement of what could be done or made, also has a base in one of the more disagreeable but ubiquitous qualities of man, greed.

Science and technology refuse to remain in the separate domains to which we might like to assign to them. Let me give you an example. In 1824 a Frenchman, Sadi Carnot, one of the great geniuses of the nineteenth century but unrecognized in his time, published a monograph, "The Motive Power of Heat." Carnot describes in considerable detail his purpose in undertaking the study. He was distressed by the ascendancy of the British over the French in war and in industry and believed that the British success followed from the relatively high efficiency of British steam engines. Carnot set out to discover and to a considerable degree succeeded in discovering the basic laws that determine the efficiency of ideal steam engines. Carnot thus became the father of modern thermodynamics, and his name has an honored place in the history of science. My point is that technological need does sometimes generate splendid science and that the precise differentiation between science and technology is meaningless.

In limiting my discussion to science, I am not constrained by the conventions of the academy that create schools of technology, schools of medicine, or departments of science. An essential feature of science is the critical observation of the natural world, a world that extends all the way from the boundaries of the universe (if they exist) to the most minute entities whose existence we have been able to validate. The observed world extends from the gross structure and behavior of whales to the genetic material that makes whales whales. It extends from the process of formation of continents to the microscopic structure of ocean sediments. I include in the process of observation the contriving of devices and procedures for enlarging our capacity to observe nature in great detail and with precision and also for extending the range of our senses to distances, masses, and intervals of time that our unaided biological senses cannot encompass. The process of observation is a creative art. Nature must be subtly questioned to reveal its secrets. To turn from art to very mundane matters, it is observation that costs money. As we try to extend our observations to more minute entities of nature, to more remote entities, or to more fragile ones, the devices that allow observation become increasingly complex and costly.

Science would not exist without the discovery of relationships between events. Such discovery is not casually made. It requires imagination and insight. The triumph of science is the building of inclusive intellectual constructs that describe the behavior of nature with elegance and precision. This is the highest creative act in science, and it marks the greatest names in the history of science.

Science is done in response to the inner imperatives within the scientist, the need to know. A scientist has a sense of incompleteness in being unable to understand the detailed interrelationships between the phenomena of nature. That sense of incompleteness may occur in crucial matters that are fundamental to our groping for the grand scheme of the universe or in matters that are modest details within the grand scheme. The compulsion of wanting to know can be strong and urgent, even if an answer is sought to a minor question.

I do not want to suggest that the scientist is not subject to any or all of the common human motivations. He enjoys success, fame, and the rewards of fame. He likes to believe that he has achieved a very small toehold on immortality. Nevertheless, the basic urge is to understand.

I recognize the immense impact of science on the style and quality of modern life through its contribution to the creation of technology. The relationship between science and technology will persist. Technology both deserves and needs the best that science can offer.

It seems clear that technology is here to stay. There is simply no possible return to a romantically perceived Arcadia; indeed, a realistic appraisal of the past suggests that its Arcadian qualities were very small. Its rustic simplicities would, I suppose, be charming could they be superimposed on the often grace-giving products of a sophisticated technology. It is, however, irrational to suppose that simplicity and sophistication can effectively co-exist in something as complex as the sheer num-bers of people require society to be. I have an immense respect for technology, although I agree that there are excesses of technology, that valuable technologies almost inevitably have undesirable side-effects, sometimes ineradicable, and that there are technologies with very few redeeming social values. Still, I am pleased to live in a world blessed with technologies that have eased and enriched the process of living.

The future of science as an intellectual pursuit obviously depends on the future of man and of his society. It has, for almost three decades, been conceivable that man will totally destroy his own kind and much of other life. Today the capacity for total destruction exists, and the constraints against destruc-tion are not wholly reassuring. Nevil Shute in his book of *On the Beach*, published in 1957 (1), has written a plausible scenario for man's self-destruction. The self-destruction is not a willful act: an ill-considered and uninformed response to a foolish act, damaging locally but not intrinsically of worldwide impact, unleashes the machinery of total destruction.

If the stream of human life ends, science will also end. Man's organized cognizance of the world is essential to the existence of science. If life on this planet were destroyed, the universe would relentlessly continue to operate in its own exquisite way, but knowledge of this would be forever denied to man.

Aldous Huxley in *Ape and Essence*, published in 1948 (2), also predicates a vast, almost worldwide nuclear catastrophe but one that falls short of the annihilation of man. Civilization as we know it is destroyed in all but a small part of the globe. The setting is in Los Angeles many generations after the catas-trophe. A symbol of the destruction of the record of man's acquired knowledge is the burning, as fuel, of the contents of the Los Angeles Public Library. Still, as long as some remnant of humanity survives, as in Huxley's tale, the extraordinary quality of curiosity that leads to science will also have managed to survive. It seems probable to me that a new civilization would, in the postulated circumstances, grow slowly, together with a new science. It could even be that, after perhaps several millennia, a new Newton would address himself to the same problems as Sir Isaac did and find the same exhilaration as Sir Isaac in their resolution.

The detailed eventualities that have been described by Shute and Huxley are, perhaps, remote. I think, however, that we will not long persist in a system of social organization that is an extrapolation or a measured evolution of the present one. To anticipate how any human activity will fare in an exceedingly precarious future is impossible.

The present social, economic, political, and military instabilities are not conducive to the cultivation of critically examined knowledge. What seems to be called for is action of some kind to meet clear problems — poverty, hunger, pollution and other environmental damage, diminishing fossil fuel resources, prejudice, and the breakdown of social order. Action, sometimes ill-considered, takes precedence over thought. An example of this is very much on my mind. America's universities are increasingly turning to vocationalism, which is not the same thing as education in which a critical and informed mind is cultivated. Increasingly, the university sees itself as a collectivity of problem solvers; although problems can be solved only if there is knowledge, their solution may offer very little new knowledge.

Learning has, in this century, been officially suppressed in nations with a magnificent civilized and civilizing tradition during periods of immense social and political ferment. The burning of the celebrated Library of Alexandria in the seventh century has been followed by a good many other book burnings, some symbolic of the suppression of learning and some real attempts to destroy the record of learning. It is not inconceivable that learning may again be suppressed.

I do not think that the record of the knowledge described as science can be wholly destroyed in any short time. Thanks to the invention of printing, the record is too widely dispersed over the earth to make its total destruction possible in any conceivable set of circumstances created by man. But I must at once assert that a future of science in which its present canon remains static is no future at all. An essential quality of science is its dynamism, the continuing search for new knowledge of nature, for new insights, for new formulations of ever-increasing inclusiveness, precision, and elegance.

I believe that knowledge deteriorates if it is not continuously reexamined, amplified, reformulated, and enlarged. Unexamined knowledge ultimately becomes superstition, which has a life of its own, quite independent of its source in human experience and of its structuring by the human mind. The superstition of today is the knowledge of a former time.

In thinking about the topic of this symposium, I found it useful to imagine what might have been concluded at earlier symposia on the same topic. For the purposes of speculation about science as a whole, I have drawn on that part of it I know best, physics.

Many of the great classical physicists of the nineteenth century are reputed to have believed that the future of physics lay in the next decimal point in the measurement of physical quantities. The assertion described a common but not universal belief that all of the fundamental laws of physics had been discovered. To be sure, there were conceded to be modest gaps in knowledge; as an example, it was not self-evident how the canon of classical physics was to be applied to the interpretation of certain phenomena that had recently been discovered. The view of the completeness of the picture of natural phenomena and their interplay with theory did have merit. Almost everything seemed to fit very precisely into a mechanical conception of the universe.

The luminaries of the time, attending a symposium on the future of physics near the end of the nineteenth century, say

1890, would not, I think, have even begun to anticipate the revolution in physics that was about to occur. In terms of the known natural phenomena that had been explained, the unexplained, known phenomena were few. The participants in the symposium could not have known that the unexplained, known phenomena were central to the creation of a new physics of a different kind. The wholly detached observer in classical physics became an integral part of the system under observation in the new physics. They would, almost certainly, have predicted an interesting future, but hardly one as exciting as has been the more-or-less immediate past.

The first quarter of the twentieth century was to produce wholly novel intellectual frameworks within which physical phenomena were interpreted. These frameworks included both relativity and quantum theory. New experiments led to new postulates, of which certain classical postulates were only limiting cases. Evolving theory, in turn, led to inspired experiment. It was an exciting period for physicists, and the excitement formed thought far beyond the realm of physics.

If another symposium on the future of physics had been held in, say, 1930, the participants would have recognized the transforming advancements in recent decades. They would not have asserted or even suggested that some kind of ultimate and all-inclusive truth had been found. I suspect that the symposium participants would have learned, perhaps with delight, that the qualities of being ultimate and all-inclusive may be only transient properties of perceived truth. New and exciting phenomena were being observed at a rapid rate. As an example, the neutron was not to be discovered till 1932, but unexplained phenomena observed in 1930 clearly led to experiments that convincingly demonstrated the existence of the neutron. A thoughtful symposiast would have exercised considerable caution in any detailed prediction of the future of physics, but the general prediction would certainly have been one of great optimism.

It seems to me to be unlikely that, in any public discussion of the future of science prior to World War II, the funding of science would have had an important place. The principal source of support for scientific research came from the universities that paid the salaries of teacher-scientists. The budgets of both public and private universities contained very modest lines for the purchase of equipment and materials, as well as for the salaries of machinists and other technicians. It was assumed in both the public and private institutions that scientific research was a normal function of members of the scientific faculty. I suspect that the identifiable budget for research was not large enough to trouble the trustees of private institutions or to become a political issue in the public institutions. In 1949 I became Chairman of the Department of Physics at Columbia University and looked at old budgets. In the late 1930s the annual budget for research was $15,000, not including salaries for machinists. Yet both Fermi and Rabi were members of the faculty and did most productive work. Some few foundations made grants, very modest by today's standards but also of critical importance in the support of research.

A few privately endowed institutions dedicated all or part of their resources to scientific research. Two such institutions come to mind — The American Museum of Natural History and the Carnegie Institution in Washington. Governmental agencies, such as the National Bureau of Standards and the Naval Observatory, did important research. Finally, men at some few industrial laboratories made fundamental contributions to basic science. The totality of universities was, however, the major contributor to an evolving science.

An intense concern with money did not loom large in the laboratories of America. To be sure, I have no certain knowledge of this. Before World War II, Wheeler Loomis, John Tate, I. I. Rabi, and George Pegram worried about these things on my behalf. Still, I do know what was preoccupying Rabi's mind, and its unit was not the dollar.

Physics, and perhaps all other sciences, was recognized for what it was — an inquiry into the way in which nature works.

An ingenuous amateur in science might have asked what you had discovered since he last saw you, but he was not likely to inquire what you had recently done to alleviate the unhappy human condition.

I do not, of course, know how great scientific productivity before World War II might have been had science then been the beneficiary of vast federal largesse. I do know that science was exciting, that there was an effervescent mood among its practitioners. All of us will end our lives with the certainty that potential knowledge now unknown will never be known to us. We will inevitably miss something, but we do not know what. Nevertheless, if in our lives a former ignorance becomes light, we have achieved an important purpose. I think that the process of acquiring understanding rather than the rate of its acquisition is the great experience.

Austerity in the practice of science may, perhaps, stimulate imagination. Let me give you an example. The properties of atomic nuclei, especially their spins and magnetic dipole moments, were essential data in the development of nuclear theory before the war. One extremely important source of information was the optical hyperfine structure of atoms. With virtually unlimited resources, it would have been reasonable to try to stretch the state of the art of classical optics to extend both the range and the precision of measurement of hyperfine structure. With the invention by Rabi of the poor man's method of measuring nuclear magnetic moments in 1937, a technique known as the molecular beam magnetic resonance method, the precision of measurement was increased by many factors of ten; the desired result was found from very nearly raw data, whereas the result as found from hyperfine structure required rather subtle theory (3). Further, Rabi's original method evolved to allow a whole range of study of atoms and molecules not even contemplated before 1937. It is certain that sheer poverty was not the motivating impulse to discovery; still, it did not thwart it.

I do not think a symposium on the future of science could

have been held during World War II. It is true that both science
and scientists served the military needs of warring nations. The
essentially reflective nature of science is, however, antipodal to
the spirit of war. Science is the output of the universal human
spirit, war the end point of an irrational parochialism.

World War II provided for science an unprecedented aura of
success within which the future of science could be plotted. The
war convincingly demonstrated the immense power of the
body of knowledge of physics; it also demonstrated the utility
of the framework of ideas within which the physicist works. I
think that no other science was as highly regarded as physics
immediately after the war. The awe of physics arose largely
through the production of the nuclear bomb. The abstractions
of physics, the intellectual toys of the physicist suddenly became immense forces in both the prevailing world structure and
in any conceivable future world structure. The support of research in physics rose very rapidly after the war in response to
the demonstrable success of physics and in anticipation of its
continued success. I use the word "success" in the utilitarian
sense.

Large-scale Federal support of research in the universities
came from the United States armed services. Initially, the war-
time mentality of mission-oriented research with the attendant
classification was common. This was rapidly modified, and
university research sponsored by the Department of Defense
became largely unclassified. In fact, research scientists were
generally uninhibited in the choice of and approach to problems. The record of achievement in the post-war period in
laboratories supported by the Department of Defense is impressive; the achievements will remain a part of man's intellectual heritage long after the source of funds is forgotten. Of all
the large, mission-oriented agencies of government, I mention
only the Department of Defense, because of my own detailed
experience with it. I would describe the style of the men from
Washington who administered research grants as one that
would grace the most enlightened academic administration.

The National Science Foundation was not mission-oriented in the sense in which the Department of Defense and the Atomic Energy Commission were, though it was subject to political pressure. Again, in the heyday of support for scientific research, it played an important role.

On a somewhat different time scale, all the sciences became beneficiaries of federal largesse. At some future time, when the enduring significance of scientific research can be more objectively examined than now, the record will almost certainly show that the critical experiments that were done and the important ideas that were developed between the end of World War II and, say, 1975, owe a great deal to Federal support.

If a symposium with the title "The Future of Science" had been held in 1955, or even in 1965, when I was a participant in the first of these Nobel Conferences, the almost certain conclusion of the symposium would have been that the future of science is limitless. We could not have stated our conclusion more eloquently than by a quotation from Kepler in 1596. "Just as nature sees to it that the living will not lack victuals, so we may say justly that the diversity in the phenomena of nature is so great and the treasures hidden in the heavens so rich, precisely in order that the human mind shall never be lacking in fresh nourishment, in order that man become not satiated with the old nor stay at rest, but rather that he find the world an ever open workshop for matching his wits (4)." We might, perhaps, have adopted a resolution of appreciation to the statesmen-politicians-scientists among us who managed to persuade governmental agencies to finance rapidly growing science, and, indeed, to create a new institution dedicated to the support of science. The most eloquent among us would, on our behalf, have written a few sentences applauding the spirit and style of that support.

We would have recognized the enormous formative influence of science-created technology on human life. We would have concluded that, on the whole, the trend of human life was one of continual improvement thanks to a highly sophisticated

technology. To be sure, we would have recognized the potential
hazards of ill-considered technology. On the other hand, we
would have expressed a considerable faith in the rationality of
man; that is, we would have stated the belief that sheer enlightened self-interest, both for the individual and for the society of which he is a part, would lead to wisdom in the application of science and a derivative technology to the affairs of
men.

Things have changed in the last decade. Mankind is increasingly concerned with its own future. We are obsessed with real,
not imaginary, problems. We are informed about them, sometimes in a partisan spirit, through magazines, books, television, radio, movies, the theatre, and the pulpit. Even a brief list
is staggering: increasing world population; malnutrition and
starvation in much of the globe; pollution of our air, waters,
and land; our incapacity to meet the ever-growing demand for
energy over any considerable length of time; the hazards of
radioactive contamination consequent to a nuclear technology;
the depletion of what once appeared to be boundless natural
resources; the decay of cities, once the glory of civilized man. I
feel compelled to add to this list of woes of modern man the
problem of equity among men, a problem as old as civilization.
None of the problems are new; each of them has passed the
critical historical moment when the concern of a few has become a social awareness. Most of the problems have an important technological component; that is to say, informed persons
believe that technology may at least alleviate the problems or
contribute to their solution.

We also live in a time of gross economic dislocation, not
unrelated to the problems I briefly mentioned. Each of us feels
threatened by a most uncertain economic future. In a burgeoning society with prospects of a generous future, we would
cheerfully sanction and even applaud increased social expense
for things that illuminate and decorate the life of man. Under
present circumstances, however, there is a considerable resis-
tance to the allocation of governmental money to purposes that

are deemed to be relatively unimportant to the present world.

Policy in all the Federal granting agencies is increasingly to support applied research rather than pure research, that is, to support the development of technologies presumptively useful in solving our problems, rather than to an inquiry into the unknown. Scientists everywhere are said to be distressed by the unavailability of support for their work.

The residual funds for science are, of course, very great. We are about to send a spaceship to Mars with the specific mission of determining whether living matter occurs on Mars, or ever did. The implications of a definitive answer are enormous. The National Accelerator Laboratory continues to operate productively in the eternal quest for knowledge of the ultimate nature of matter. It would be easy to make a long list of major research installations that continue to operate, although perhaps on diminishing budgets as measured in dollars of fixed value. Without having undertaken the very difficult task of estimating the quality of the work that is still supported in style, I nevertheless believe that considerable intellectual taste has been used in the choice of work to be supported.

A scientist friend of mine remarked to me during the lavish phase of science that what physics needs is a good depression. Another friend has remarked that there are needlessly many physicists who are described as high-energy theorists. Neither of these men make judgments, even private ones, that are ill-considered. Almost every physicist I know privately agrees that the literature of physics has grown far more rapidly than its content of ideas. The great tradition of physics centered around ideas; ideas went in search of money and also in search of techniques or technologies to allow the exploration of ideas. In the lush period of physics, both money and a clever technology went in search of ideas. These ideas did not, I think, contribute as much to the essence of science as the ideas that searched for money and technology.

The last decade has been marked by major discoveries in many branches of science. The discoveries have raised more

questions than they have answered. We are very far from the
point at which we might conceivably lean back with satisfaction in what we have learned and assert that the conceptual
structures that we have created to describe observed reality are
some final truth. In fact, there is a great deal of elegantly
observed reality that does not fit into any organized set of
ideas. The challenge for a further growth of science is there; the
momentum of the acquisition of knowledge is high. I see
nothing in the offing that would suggest a future for science less
brilliant than its past. I exclude, of course, the eventualities I
discussed earlier — the destruction of man or the destruction of
civilization.

I think that it is one of the fundamental rights of man to
inquire, to observe, to think, and to analyze. I think that that
right is as fundamental as any other asserted right. As for any
other human right, the exercise of the right to inquire must not
infringe on the rights of others. It is not a corollary of these
statements that every putative inquirer into nature should
have, or must have by some fundamental right, the virtually
unlimited support of the society as a whole. Freedom of inquiry
does not give the support of science an intrinsic priority over
the support of a large number of other civilized and civilizing
human activities. For example, historical or linguistic research,
the writing of novels, the composing and playing of music, the
writing and performance of plays, the making of sculpture do
not self-evidently rank below science in their proper demand
on the public purse.

We cherish the notion of a democratic society and are, I
think, committed to a strengthening of our democratic institutions. The judgment about the degree of the public support of
science in the face of other demands on the national wealth
must be made by representatives of a group much larger than
that of the scientists. If science is to be generously supported,
that larger group must be persuaded of the value of science. The
value has several aspects, one of which is the quality of science

to which my remarks have been directed, that is, science as a magnificent adventure of the mind and spirit of man.

We, the scientists, find pleasure in scientific knowledge. We are exhilarated when we learn of a new discovery or of a new formulation of science; we are exalted when the discovery is our own. If our constituency shares, in a modest way, the pleasure and stimulation we find in science, the satisfaction in understanding the order that pervades the universe, the future of science would be more certain.

A number of periodicals directed to the more contemplative segment of society, much larger than the scientific community, print admirable articles about science in the present sense of the word. I think that these have a dedicated audience.

We never have learned the knack of explicating the purpose, spirit, and style of science to students in the universities and colleges of America. Otherwise well-educated students leave the university totally unknowing of any part of the scientific tradition and uncomprehending of the extraordinary vitality that pervades modern science.

Many of them seem to view science as advanced plumbing or arcane and damaging magic. I have a feeling in talking to some of those from disciplines outside of the sciences that I may be asked to repair their television sets; with others I have the feeling that I personally have despoiled Lake Erie, intend to uproot the State of Wyoming in the search for coal and am otherwise wholly free of the values of a humane and civilized man.

It may be that the scientific fraternity itself creates a mood in which science is dull, obsessed with trivia and without heroic vision. Nothing gains stature and the rewards of stature as much as research. Research may be only peripherally related to any major intellectual question, and sometimes it has no significance at all except to produce a missing number in tabulations of experimental data. I do not intend to challenge the value of such research work to a teacher-scientist; it is a connection to the adventure of science if not its substance. However, the teacher-scientist may become so immersed in the details of his work that he displays no suggestion to his students of the adventure that is inherent in science. Why should not all educated persons find pride and pleasure in the achievements of science?

Finally, the future of science depends on an understanding by the people at large of the relationship between science and technology. There can be no doubt that the present body of scientific knowledge can serve as a base for a vastly more sophisticated technology than we now have. Still, much innovative technology is the consequence of relatively new knowledge, of what were only recently the abstractions of the scientist. While science provides the knowledge that allows the development of a technology, the motive energy for creation and use of a possible technology has many sources. These sources include human needs — whether real or imagined — the economics of the technology, and one's hope for the future of man. The support of science does not, of necessity, produce new technology, whether valuable or damaging; the support of science will yield knowledge that offers mankind large options as the future is planned. Again, these are matters of which undergraduates in general gain no knowledge or insight. In fact, students of science have not ordinarily concerned themselves with these matters. We could, perhaps, do better by our students than we have done. Man could better move into a difficult future if the present and potential options offered by science were better understood than they now are by those who give leadership to our society.

The future of science cannot be separated from the future of civilization. If my hopes for the future of mankind and his civilization are realized, then I am confident of a bright future for science.

REFERENCES

1. Nevil Shute, *On the Beach*, Morrow, 1957.
2. Aldous Huxley, *Ape and Essence*, Harper, 1948.
3. I. I. Rabi, J. R. Zacharias, S. Millman, and P. Kusch. "A New Method of Measuring Nuclear Magnetic Moment," *The Physical Review*, **53**:4, 1938.
4. Johannes Kepler, in Werner Heisenberg, *The Physicist's Conception of Nature*. Translated by Arnold J. Pomerans. Greenwood Press, 1958.

DISCUSSION

General Comments

Dr. Segrè. I'd like to comment on something Dr. Kusch said in his speech. He said, "If a symposium on the future of physics 'had been held in 1930 the participants would have recognized the advancement in recent decades." Now, there was a symposium in 1929 and at this symposium there was a speech given by Professor Corbino who was a leading physicist in Italy at that time. This speech was given by Corbino, but was written together with Fermi, and it said different things from what Dr. Kusch imagined they would have said.

Corbino first described what future experimental physicists would do. First, he explained how physics operated. Some physicists searched for new phenomena, like discovering the electric current, discovering X-rays, or discovering radioactivity, discoveries which were God-given in a certain sense and which nobody could predict. Then there were those who measured things with always increasing precision. It may have been dull work, but it was important work, nonetheless. Then there were people who did something intermediate. They tried to draw consequences out of theories and tested the validity of these theories.

The second part of the speech was a prediction of what would happen next. He said atomic physics had reached a stage of maturity, and that there would be no great surprises. No new electromagnetic radiations would be discovered because the whole field of electromagnetic radiations had been covered. He

---

Editor's Note. Asterisks denote a break in sequence or a change from one discussion panel to another.

said, I think, that atomic physics, molecular physics, had come to an end in the sense of great new fundamental discoveries.

But then, where were the new discoveries to come from? Well, there was still an unexamined nook, and that was solid state. He said that if one would succeed in purifying substances to an extreme degree, something new and unforeseen would happen. But the real future would be the study of the nucleus. The nucleus would be attacked and the mysteries one surmised in the disintegrations observed by Rutherford would be solved by building accelerators. Of course, in 1929 there were no accelerators, but he said that the technology was more or less ripe and that it was only a matter of money and of making them. This would lead to the great goal of liberating atomic energy. Now remember, the neutron had not yet been discovered, but nonetheless they still saw that in the future.

A final prediction was that modern physics would be applied to biology. This would be done, not by combining the physicist and the biologist, but by putting the physics and the biology in the same brain so the same man could do it. That's the end of the speech, but it remains a pretty good prediction of what actually happened.*

* * *

Dr. Beadle. As a biologist I couldn't help comparing Dr. Kusch's presentation of physics, the development of it, the excitement of it, and, to some extent, the use of it, with my own field. There is much similarity in development; much of the excitement is shared by both fields and, of course, other fields of science. But, he quite properly limited himself mainly to the development of physics, the excitement of physics, and said a good deal less about the practical uses of physics and in general the misuse of science. The opportunities for use and misuse of

---

* (This speech is to be found in *Atti Soc, Italiana Progresso de la Scienze*, 1929, **18**, p. 1157.)

science appear across the board, of course. We can misuse physics; we can misuse chemistry; we can misuse biology. The scientist by and large is concerned with developing science, developing what knowledge is available about the various areas and often is less concerned with the misuse. The danger of continued misuse is a serious threat to the whole of humanity. An important question is, "How can this threat be lessened?"

It can be lessened only by a much wider understanding of science on the part of large populations. What we need somehow in our educational system is a way of making the general population aware of the dangers of the misuse of science. How we should do this I don't know, except that it has to be an understanding on a worldwide basis as to the damage that can be brought about by the misuse of science. I want to emphasize that point without in any way distracting from the usefulness and the benefits that science has given us.

*   *   *

Dr. Onsager.   Dr. Kusch mentioned that ideas in search of a technology are much better than technology in search of ideas. I'm not quite convinced of that. I think it's very much a give and take. We might remember what Lavoisier did for chemistry when he introduced a systematic application of the technique of weighing, or Bunsen, when he examined flaming substances with a spectroscope. This discovery not only gave the chemist a very incisive qualitative tool for the identification of elements, but in due time led to the discovery of systematics in spectra and other fundamental ideas related to light.

The invention of the mass spectroscope was another technique which allowed chemists to distinguish between atoms which differed very little except for weight. These discoveries, in turn, gave a good deal of insight into nuclear physics. Accurate measurements also provided the basis for (not denying the importance of Einstein's identification of energy and mass) an accurate prediction of the amount of energy to be derived from a nuclear reaction.

A last example is the development of very accurate magnetometers using the nuclear magnetic resonance of hydrogen. Again, we see technology plus an idea. Maurice Ewing decided to systematically explore what small variations there might be in magnetic field intensities in the ocean. It turned out that there are lineations of alternating intensities which tied in with other ideas about continental drift and changes in the polarity of the earth's magnetic field. But the system according to which these magnetizations vary over the ocean floor will be discovered only because of these very accurate techniques for measuring magnetic fields.

*   *   *

Question.   The first question of the afternoon will address itself to the distinction between science and technology as spelled out by Dr. Kusch. How do you see this distinction between science and technology?

Dr. Cournand.   I'd like to paraphrase a definition of science that was given by Karl Popper (a philosopher of science). Science is an institution based on tradition, on past and present members, and on accumulated knowledge, which is devoted to the understanding of nature (environment) and of ourselves. This involves, therefore, the physical and life sciences, which are devoted to the understanding of nature, and the behavioral sciences, psychology, and sociology, which are devoted to the understanding of ourselves and of our relations with the natural and social environments. Now I have used the word "understanding" and not used the word "truth." There's only a relative truth which is dependent upon the state of our knowledge at a given time. There is no such thing as absolute truth, at least for those who deal with science.

Now, insofar as technology is concerned, I would say that technology is the application of parts or the whole of the truths, and is used in order to facilitate acquisition of further knowledge and improve life in our environment. In many instances, if you study Greek history, technology has existed independent of science. It was based on a tradition and has served as the original basis of science. Technology in the modern world has taken a very important place in the development of science. But, what I wish to emphasize is that one facilitates the other, and there is reciprocity of service between the two. In brief, a neat distinction must be made in your mind between science and technology.

Dr. Lamb.   I think my own work has been very free of practical application. I came close to practical applications a few times, but I never had the insight to make anything of them.

During the war I worked on high-power radio oscillators for very high frequencies called magnetons. I remember once a very important signal corps general was coming to visit the laboratory and we arranged to have a demonstration of the effectiveness of our research. So we focused a beam of this radiation on an area of the door through which we thought he would enter so that when he came in, he would find he had a very warm spot in his middle regions because we were illuminating him with microwave radiations. Of course, this is the way your microwave oven works.

In another experiment with those microwaves we needed to bounce them around uniformly inside a large box. We found that a large metal fan would mix them up quite effectively. This same type of fan is found in every microwave oven that you buy so that your roast will get done more evenly. These are examples of how the pure research which I was doing, even though it seemed like war research, was, in fact, pure research. I just missed practical application of this work.

Dr. Weller. I think our views here reflect our very different backgrounds, and whereas Dr. Lamb might say to you he never tackled a problem he thought would have a practical application right away, I could almost say the reverse. I've probably never tackled a problem where I wasn't seeking some answer that I thought would improve some health problem that mankind is subject to. I do get very irritated to hear statements from some scientists who say that pure science done only to satisfy one's own intellectual curiosity is the only good science there is. That I disagree with very, very much.

* * *

Dr. Gilkey. I was impressed with the rather neat way in which Dr. Kusch showed that he was unwilling to cut the umbilical cord between science and technology by insisting that a precise distinction between the two is meaningless. On the other hand, however, he made it quite clear that the attraction, the ecstasy, the glory, and the wonder of science was the desire to know, and a really quite uninterested desire to know. I thought this was quite a remarkable way of saying both things about science.

* * *

Dr. Harvey. I'm interested in what you might call the sociological conditions under which something like science comes to dominate a culture such as our own, and I tend to see the concept "science" as an abstraction for a whole set of very concrete professions and institutions. What's happened, historically, it seems to me, is that pure science has gotten linked to technology. It is this linking of technology and science that accounts for the incredible burgeoning of science, which in turn is related to the amount of money that has been poured into it, and the prestige it enjoys among young people.

There are all these conditions now that give science the place in our culture that theology might have had in the fourteenth, fifteenth, and sixteenth centuries. If that's true, then it is a little artificial to separate quite as much as some would, technology and science. Where the government and where the universities put their money and their prestige seems to me to have a tremendous effect upon the development of ideas. This raises the whole question that Dr. Kusch raised in his lecture, of what stake the public has in where they put their money. It's because of the linking of science and technology that the political and social questions arise — what are we going to do with this knowledge and this technology?

                                *   *   *

Question.   Dr. Kusch made the general assertion that at least certain aspects of science are similar to creative art. Is science a form of art or are there some things which finally distinguish the scientific from what we might normally call art?

Dr. Eccles.   What science and art have in common is that they are both the result of creative imagination. We have this extraordinary ability in our own mind to go beyond the immediate knowledge and understanding and get some new insight. And this new insight is, of course, the scientific hypothesis which can explain numerous phenomena with understanding which can be tested. And in art there's, again, a new insight, some aesthetic experience or some creative urge that comes and guides the artist. So, there's this much in common.

Then another point is that science is a part of literature. Not only do you have the experiments and the ideas, but you have to write them up in logical form, in very intelligible language, and very clearly so that they can be criticized. This is a real art form. When I'm writing scientific papers I feel the same as I think a poet feels when he's writing poetry. We have our thoughts, and we're trying to enclose them in language in the most logical way.

Dr. von Euler. One could also add that it is quite common to talk about the beauty of a discovery. I think it's quite true that certain findings have an intrinsic beauty. They fall in place or they are shaped mentally in such a way that they give very special satisfaction. I believe this is one of the ties with the arts.

Dr. Dean. I think the emphasis on the aesthetic dimension in science is helpful for dealing with the problem of the distinction between pure science and applied science. I think you could make the argument that science to the extent that it's motivated by aesthetic reasons may be pure science. To the degree that it's motivated by practical or utilitarian reasons, that may be what we mean by applied science. But I think we could go a step further. If pure science is thought of as an aesthetic enterprise, then it would seem to me to follow that pure science should be considered on a par with other aesthetic enterprises. If we were to characterize it that way, why shouldn't pure science then be funded as the arts are funded?

* * *

Dr. Barbour. I have a great deal of respect for both the role of the pursuit of truth and the social role in the motivation of the scientist, but I wonder if one doesn't, to be realistic, have to put greater stress on other factors at both the individual and the social levels.

In terms of individual motivation pursuit of disinterested truth is important, and we need to get this across to our students. But surely, professional recognition deserves more than a passing footnote if one is looking at the actual day-to-day motivation of a scientist. How many papers would get written if publications had to be anonymous? It seems to me that to be realistic one has to say that scientists are human, and like anyone else, both professional recognition and public recognition are very central motivating factors.

* * *

Funding

Dr. Hofstadter. I think Dr. Kusch perhaps left the audience with the feeling that in the good old days research was cheap and fantastically rewarding, and that these days research costs a lot of money but one doesn't get as much from that money. It's not as cost-effective, as they say nowadays.

I've been a participant in both little science and big science, if you want to use those words. In fact, I worked in the same laboratory as Dr. Kusch in the 1930s, and I experienced some of the same pleasures which he talked about. But I've also been involved in what you might call big science. I'm in it right now, as big as it can be. It's very expensive, but there isn't any other way to do it. If you're going to get beautiful, new, interesting, and potentially useful results, you've got to pay the price, because that is the only way you can do such experiments. The public, of course, can decide whether it wants to spend all this money or not. That's up to the public, and the scientists will go along with whatever the public desires.

Dr. Kusch. I didn't intend to disparage big science. I think I applauded the taste of granting agencies in supporting the National Accelerator Laboratory as one example. I also applauded the attempts to discover whether or not there was life on Mars, certainly the most expensive experiment in history.

Dr. Huggins.   I'd like to go from the sublime to the less sublime. Biology is a cottage industry. To think of spending
$100,000 on an experiment would make every biologist in the
world faint or die. In biology nothing is known about nothing.
In our field there is no theoretical biology, it's all experimental.
In experimental biology, with blood on my hands I have a
chance. Seated at my desk I have no chance. I've thought a lot
about the critical mass for discovery in biology and it's something more than one and less than three. The additional hands
around a laboratory are students taking notes.

* * *

Dr. Kuznets.   One question in Dr. Kusch's paper that has not
been answered is: on what basis does society decide how much
in the way of resources to allocate to science, even if it's the
finest, pure science? If pure science is an important part of our
life, so are many other things in our civilization — poetry,
music, certain aspects of technology. How do we divide the pie,
so to speak. As far as economic analysis is concerned, there has
never been any possibility of defining this share in any sort of
determinate fashion. Unless we can deal with this question in
some tangible and defensible terms there is a doubt that we can
arrive at firm judgments. This means, in essence, that we can
have instinctive reflections or reactions, but our judgments will
be hanging in the air.

   Historical experience is not much of a guide because in some
cases a country may have decided to rely on science from
elsewhere. The United States provided very little support for
pure science in the 19th century and there was very little interest. We had tremendous progress in technology, but we had
practically no great pure scientists with one or two exceptions
until the 20th century. Or a country may decide that science is
important for its development if it's a backward country. Germany did this in the 19th century, and built up science in its

universities with the express purpose of providing some basis for future economic growth. But how much you could invest in that, how you can calculate any kind of reasonable set of weights, is a question which cannot be fully answered.

* * *

Question.   What is the role of a scientifically informed citizenry with regard to the allocation of resources to the sciences?

Dr. von Euler.   I believe it might be difficult, really, to count on a scientifically informed community. It seems to me that sometimes it's difficult enough to make one's fellow workers and fellow scientists really understand and appreciate what's being done. Think how much more difficult it would be to extensively inform the community. So, I believe a community has to rely on the scientists to a certain extent, and I think they still could do that with confidence.

Dr. Anfinsen.   I think the allocation of money for something like an accelerator, for example, which requires enormous sums of money must almost be done without any participation by the public. It's hard for me to visualize that the public would appreciate why anybody would want to know about the properties of the elementary nuclear particles. The allocation for this type of research must almost have to be done by people who think they know what's important. I can see allocating money for arthritis, because Aunt Susie has arthritis. But, not being a physicist, it's hard for me to see how they get all the money they do.

Dr. Schwinger.   I agree that it is difficult to convey to the public at large why this particle accelerator is desirable, but I do believe it is possible and, in fact, mandatory to convey to the public at large the thrill of this hunt, the intellectual and, to some extent, the practical importance of it. In a sense our whole life style has been set by the technological developments that have flowed from the development of physics and other related sciences, and this is just the latest version of it. Surely a feeling of the mystery and the awe of nature that goes with this can be transmitted. The details are not important. It's a general feeling for the search — its importance. The fascination surely is transmittable.

Question. Dr. Kusch asserts a recent shift of emphasis from basic to applied research. Do you agree, and if so, what are the consequences?

Dr. Anfinsen. It's certainly true in the medical sciences. The cancer act that was passed is the best example of the change of thinking. We feel this very much at the NIH now. We feel this slow, creeping semiparalysis coming about through the creation of alternatives to the peer review system, for instance. Instead of one's peers reviewing one's application for grants, it's now frequently done by a contract mechanism, almost by edict by the directors of certain institutes. That's a very dangerous business, because they're funding primarily what they think might be the most profitable.

Dr. Eccles. I agree completely with what Dr. Anfinsen said. I think we have quite a dangerous situation where bureaucrats like to think they're in control of the money and that scientists are their paid servants, and with the best will in the world they try to define the kind of problems that are important for the future. This is doomed to be very frustrating to the scientist. Furthermore, and this is more important, it subsidizes immorality amongst the scientists, because those who want money will pretend to be doing something which is in line with the contract, but will manage to get something else done instead.

What is not understood by the funding agencies is that the whole future of science, the growing points, the peaks of science, are dependent upon a very few individuals with creative imagination who can see and probe into the future. The rest will follow them. Now, these are not the people who are likely to be recognized by the people controlling the contract.

Science Education

Dr. Walton.   Dr. Kusch laid a lot of blame on the teachers of science, and I suppose he had at the back of his mind the teachers of physics, for not making physics a sufficiently exciting subject to attract students to it. I don't know that this is really the case. My reason for thinking so is that the textbooks which are available to students today are far, far better in many respects than the textbooks which I experienced as a student. They are more attractively printed; the illustrations are vastly clearer and more entertaining in some cases than was the custom in years gone by. If students are not going into physics, it may be just as much the fault of the students as of the teachers. So, I would like to stand up for the teachers and not let them be condemned to the extent that they were condemned in that talk.

Now, there's this question of attracting people into science, and naturally, when speaking, I'm thinking of physics. We mustn't tell them only about the exciting things. We must also warn them in a way similar to the way in which Churchill, when he became the Prime Minister of Great Britain during a very black period in the war, warned the people that he would promise them nothing more than blood, sweat, toil, and tears. I think we have to warn people going into science in similar terms. It is hard work. Very much harder than you'd find in a great many other occupations.

*   *   *

Dr. Edelman.   It seems to me that one problem of science, at least in my personal experience, is that it's so abstract. It's not that it's really very much more complicated than what a lawyer might do, or someone in another profession, in terms of the logic involved. But, it discusses things that are very far from our senses and that are based on a number of skills that have been accumulated. It is then compressed in the textbook in such a way that it requires an enormous amount of skill before you get a feeling of it.

Whenever I've tried to explain what I do to an intelligent layman, I've always found that he nods and understands it except at the end of the visit when he'll say, "Do you mean you got the structure of that molecule from this white powder?" Then I realize I have failed. And I believe the reason I have failed is because the layman has had no opportunity to experience directly the kinds of things I take for granted as a result of 10 years of training. So, the words are really not extraordinarily useful and it takes an extraordinary metaphorical talent to translate that experience directly. That does not mean that we shouldn't try. My personal belief is that much more should be done in this direction.

* * *

Dr. Brattain.   I don't think there's any trouble with the teachers of physics or in any other science in inspiring young people. If they're good teachers, and they love their subject, then they can inspire other people to go on in that discipline. That's easy enough.

The problem arises in our society, because this discipline is so hard that a major portion of our society has no understanding of the conceptual structure that science gives. I've spent some 10 years teaching an "understanding science class" to nonscience majors. Now, you can't teach students if they don't want to come to your course, but I really feel that there should be more science in our liberal arts schools. When I went through school, you couldn't graduate without having at least one course in an experimental science class which, of course, gave you at least some understanding of the basic concepts of science.

* * *

Dr. Libby. I think that Dr. Kusch hit the college teachers pretty hard. It's been my experience that if you can get the students into the classroom, they learn some science. Our big problem is the general letters and science majors; maybe we aren't giving the right kind of course. Something like a course on the history and sociology of science might be the way to go. It is tragic to see how little science graduates of many of the colleges and universities at the A.B. level know. It's even more tragic to see how little interested they are. Now, this is a weakness and a sad condition of our society which we'll have to watch very closely.

Dr. Harvey. Dr. Kusch emphasized that it was the thrill of discovery and of achievement, of insight, that was the justification for science. If you say that, then it seems to me you are led to say, as he did, that it's just as important to have the thrill of discovery in philosophy; it's just as important to have the discovery in art; it's just as important to have the discovery in any mode of insight. No special value ought to be put on science any more than art or anything else.

The other side of this, however, is that if you say that, then I suppose you ought not to regret the fact that some students don't know any science, any more than you regret the fact that they don't know seventeenth or eighteenth century literature or something like that. But most scientists I know won't quite settle for that, and I think the reason that's so is because they really think science itself, knowledge itself, is probably the most important means of power in the modern world, that is to say, that knowledge and power are linked. This gets us back to our business of the relation of technology and science.

So I see, in this discussion, a kind of minimal confessional statement about the joys of science, on the one hand. On the other, I see a kind of suggestion that, really, science is the most important feature of man's knowledge and insight in the modern world. Therefore, you're really an impoverished human being — not only that, but probably powerless as a culture — unless you know science. Which side you come down on will give you a lot different attitude toward what education is and why science is important in education.

Dr. Cooper. I'd like to comment on some things that Dr. Harvey said. The way I would put it is, that if a person knows nothing about literature, his life is diminished, that's all. Now, he may have made that choice. If he knows nothing of art, his life is diminished. Anyone can make that choice. I feel if you know nothing of science, again your life is diminished.

I think what many of us who are in science complain about doesn't proceed from the statement that science as an intellectual entity is more valuable than these other endeavors. Rather, in general, I think it's, in fact, true that most people do know less about science than they do about literature or art. After all, almost everyone knows how to read and has even on occasion read a book, and you just have to have seen a painting or listened to a piece of music.

But it's often the case that people do not learn the rudiments of the language of science and so in effect know very little other than an accumulation of facts. They know very little about what is being done there and what the enterprise is. It's not so much that science is an intrinsically more valuable enterprise than others, but rather that people simply know less about science.

Dr. Cournand. I was quite impressed by what Dr. Kusch had to say about how very essential it is to emphasize the education of the mind, that is, not too much emphasis on new knowledge or new facts or new tools of thought. It is essential to teach these things, but it is the education of the mind that permits the student to face new situations and new knowledge and to be capable of adaptation. This is, I believe, what is fundamental to teach and it is quite different from what is usually done.

Dr. Weller. I think there's justified concern about a lack of balance between classical education and education focused on the technique of problem solving, per se. My concern extends beyond that because some institutions are now not focusing on the scientific facts needed to solve a problem, but are turning out "professional" problem solvers. There seems to be the feeling that if you mix a little introductory science in a curriculum with some economics, social science, and some political science, then you will graduate an individual who will have the capacity to make the basic decisions that are going to influence not only science but the whole development of medical care and health systems in our country. This gives me great concern.

* * *

The Brain-Mind Problem
as a Frontier of Science

by

JOHN C. ECCLES

The future of science can be dealt with as general problems: for example, growth or regression, economics, big science
versus little science, the future of biology, or space science. I
decided to devote my talk to one special aspect of the frontier of
science. Can we define the limit beyond which science may
never go? Or should go? I attempt to do this in a very special
field that has been central to my scientific life, hence my life.

Weisskopf (1) has made a valuable distinction in his concept
of the two frontiers of science. The internal frontier is a very
broad area in which the basic scientific principles are not in
question, but in which the phenomena are so complex that they
are not yet understood and explained; nevertheless, they appear to be encompassable by the present range of scientific
hypotheses and investigations at the atomic, molecular, geolog-

ical, and biological level. The external frontier borders on "those realms of nature that lie beyond presently understood principles," realms which, according to Weisskopf, comprise the nuclear and subnuclear worlds. However, I would suggest that even incomplete solutions of the brain-mind problem will entail such revolutionary changes in existing science that the external frontier is transgressed. On the contrary, Weisskopf adopts the popular parallelist philosophy on the brain-mind problem in his statement:

> Brain research is currently one of the most important internal frontiers. Difficult problems like the nature of thinking and memory are approached from two directions, through neurophysiological methods — studying the physics, chemistry, and biology of the nerve systems — and through psychological methods — studying the observed phenomena in the brain. This research can be thought of as boring a tunnel from two sides: the two approaches have not yet met but hopefully some day they will. (Reprinted by permission)

It appears that all we need is more neurophysiology and more psychology (which is curiously defined as studying the observed phenomena in the brain); hence the frontier is internal. But I think this is a mistake, and to make my objection explicit I will briefly introduce the three-world philosophy recently formulated by Popper. These three worlds comprise everything that is in existence and in experience.

World 1 is the world of physical objects and states. It comprises the whole cosmos of matter and energy and all of biology, including human brains. World 2 is the world of states of consciousness and subjective experiences. It comprises in totality our sensory perceptions and cognitive experiences, even to creative imagination at the highest level, together with the self or ego which for each of us is the basis of our unity and continuity as an experiencing being. World 2 is, initially, private to each of us; thus it is subjective; but it can become objective when it is revealed to other selves by linguistic or artistic expression or by gestures, which may be at all levels of subtlety. Finally, World 3 is what Popper calls the world of objective knowledge. It is the world of culture that was made by man and that reciprocally makes each man in his own lifetime. It comprises the whole of culture and civilization, most importantly language. World 3 includes all the expressions of human creativity that have been preserved in coded form on World 1 objects, such as the paper and ink of books.

The creative efforts to give a scientific account of nature constitute one of the most important components of World 3. Brain research is thus a World 3 activity, but the objects studied, namely brains, lie entirely within World 1. Accordingly, I would agree with Weisskopf that brain research itself can be classified as involving an internal frontier of science. There is immense complexity, but we assume that the essential principles are known. It is otherwise with the brain-mind problem, because this problem concerns interaction between two different worlds, World 1 of the brain and World 2 of the self-conscious mind. Hitherto science has been restricted to problems arising within World 1. It is evident that if the mind-brain problem is to be subjected to a scientific attack, it will be an attack across an external frontier in an extreme degree.

Most brain scientists and philosophers evade this confrontation across such a horrendous frontier by espousing some variety of psychoneural parallelism. The conscious experiences are regarded as being merely a spin-off from the neural events, every neural event being postulated by its very nature to have an associated conscious experience. This simple variety of parallelism is certainly mistaken, because the great majority of neural activities in the brain do not give rise to conscious experiences. Parallelism also is unable to account for the experience that thought can give rise to action, as in the so-called voluntary movements, which must mean that cognitive events can effect changes in the patterns of impulse discharges of cerebral neurons. An even more pervasive experience is that we can, at will, set in train neural machinery to recall conscious memories from the data banks in our brains, and then judge the correctness of the recalls.

The psychoneural identity hypothesis has been very effectively criticized on philosophical grounds by Polten (2). More recently there is a comprehensive criticism of parallelism in all its various guises in a book in the course of publication (3). Parallelists make no serious attempt to develop any scientific explanations of how all neural ·happenings give rise to conscious experiences. The only philosophy that consistently attempts to relate to this extraordinary linkage is panpsychism, which espouses the absurd notion that there is an extremely elemental consciousness even in atoms and molecules, and it becomes progressively more refined through hierarchic levels from inorganic matter to living organisms — both plants and animals — and so up to its climax in the human brain.

The most telling criticism against parellelism can be mounted against its key postulate that the happenings in the neural machinery of the brain provide *a necessary and sufficient explanation of the totality both of the performance and of the conscious experience of a human being.* For example, voluntary movement is regarded as being *completely determined* by brain events, as are all other cognitive experiences. But as Popper states in his Compton lecture (4):

> According to determinism, any theory such as say determinism is held because of a certain physical structure of the holder — perhaps of his brain. Accordingly, we are deceiving ourselves and are physically so determined as to deceive ourselves whenever we believe that there are such things as arguments or reasons which make us accept determinism. In other words, physical determinism is a theory which, if it is true, is unarguable since it must explain all our reactions, including what appear to us as beliefs based on arguments, as due to purely physical conditions. Purely physical conditions, including our physical environment make us say or accept whatever we say or accept.

This is an effective reductio ad absurdum.

With the rejection of parallelism and panpsychism, we confront the great problem of the interaction of brain and mind — the "world knot" of Schopenhauer. This interaction occurs across a frontier between the world of matter and energy (World 1) and the world of the self-conscious mind (World 2). The question arises: How far can this problem be investigated by scientific methods? Certainly, there has to be a radical extension of the scope of scientific investigation. It is difficult to put strict and final limits on what science can accomplish. For example, Weisskopf (1) states:

> The claim that science will, in the future, be able to understand every observable phenomenon, may perhaps not be wholly unjustified . . . . However, implicit in this claim to completeness is a very important qualification. If one asks the question: can, does, or will scientific insight cover every aspect of human experience, the answer must be negative. To show that this statement does not contradict the completeness claim, let me give a simple example. (Reprinted by permission)

His simple example is a Beethoven sonata, but he goes on to expand these exceptions to the completeness claim by the addition of all the problems of ethics and personal relations and aesthetics.

Bronowski (5), too, distinguishes between two modes in which human beings attempt to understand. Firstly, there is what he calls the poetic mode and, secondly, is the scientific mode. As expressed by Bronowski (6), modern physics also leads to an enlargement and a fundamental change in the relation of the human observer to the world of experience which he is trying to understand scientifically. As he states,

> These principles express a far-reaching revision in the idea of knowledge. They shift the emphasis away from the impersonal record and they put in its place a relation from which the human observer cannot be abstracted. The scrutiny of experience is no

longer idealized as an activity that could be carried out by a machine. There is no reality, there are no laws, that can be separated from the process of their discovery; the human condition is also the necessary condition for the recognition of order in the world.

We are far removed from the inductivist philosophy of science, in which the observer was thought of as collecting data and extracting therefrom the laws of science with a quite impersonal machinelike exactness. Yet this cleavage between the scientist on the one hand and the phenomena of nature on the other provided the operational conditions during the great advances of science from the seventeenth century on until the twentieth century. Scientists were deluded by the inductivist view of science, so forcibly expressed by Bacon and Mill. Though the key role of creative imagination was overlooked in theory, it was exemplified in the achievements of the great scientists, by Galileo, Newton and Darwin, for example. By contrast, I can quote from Wigner (7) with respect to such a simple happening as a scientific measurement.

> The measurement is not completed until its result enters our consciousness. This last step occurs when a correlation is established between the state of the last measuring apparatus and something which directly affects our consciousness. The last step is, at the present state of our knowledge, shrouded in the mystery and no explanation has been given for it so far in terms of quantum mechanics, or in terms of any other theory.

This quotation focuses our attention again on the brain-mind problem.

In the brain-mind problem there are, necessarily, two completely different sets of phenomena to be related. On the one side there is the study in more and more detail of the events happening in the human brain in relationship to self-consciousness. So we come to develop concepts of organized complexity and subtlety in the neuronal machinery of the cerebral cortex that are of a different order from anything known to exist elsewhere in nature. Recent investigations on patients with section of the corpus callosum (the split-brain operation) have disclosed that the area of the human cerebral cortex giving conscious experiences to the subject (the so-called liaison brain) is restricted to the dominant hemisphere (the linguistic hemisphere), and even to specialized areas of that hemisphere, which is almost always the left. There is now an impressive ensemble of experimental investigations on the details of the neuronal events that are related to conscious experiences. On the other side, there are now many examples of refined investigations within the immense range of conscious phenomena of all kinds — from simplest perceptions to the highest levels of creative imagination.

It is not in question that the happenings in the cerebral cortex are *necessary* for the experiences of consciousness by the subject. However, it must not be naively assumed that these brain events are *sufficient* for the conscious experiences, that is, that World 2 is simply a derivative of World 1. This, in fact, is the parallelist position. If we are to avoid falling into parallelism, with its self-stultifying philosophy of determinism, we must develop a dualist-interactionist philosophy, according to which the self-conscious mind has an identity and activity that are not entirely dependent on brain events, these events being under a determining and controlling influence from the self-conscious mind.

I now return to the question: How can we have confidence that we can advance scientifically in this study of the brain-mind problem? Our questioning is objectively based if we propose explanations or hypotheses that can be criticized by others in the light of known facts and existent hypotheses. Furthermore, these new hypotheses should serve as a challenge to further testing by experiments specially designed for the purpose. It is naive to expect that the hypotheses will be ultimate solutions. Rather, we would anticipate that they could provide limited and provisional explanations.

The tunnel analogy quoted from Weisskopf is too simple. It

can serve quite well in relation to a scientific attack on problems that are within World 1. However, when we are embarking on an attempt to understand scientifically problems that relate to happenings in two different worlds (World 1 and World 2), that is, across an external frontier of science, we have no guidelines to help us lay out a comprehensive and organized plan of attack on the brain-mind problem with its two fundamental aspects, as is implied in the tunnel analogy. Rather, we should try to interlock many approaches on the great problem, at the same time being very modest in our expectations. We must not expect one blinding flash of discovery. Instead, we have to dare to put up hypotheses that attempt partial explanations — for example, in conscious perception, in conscious memory or in voluntary action, where there is already an impressive assemblage of experimental testing that has been carried out in a refined and objective manner. Furthermore, since the essential theme of the problem is the human being and his conscious experiences, we can draw on a wealth of scientific evidence relating to the way in which lesions (injuries) of the brain result in changes in conscious experiences as reported by the subject or as evidenced by his behavior.

It is important now to develop an hypothesis on the mode of interaction between the self-conscious mind and the brain that is much stronger and more definitive than any hypothesis hitherto formulated in relation to what we may term the dualistic postulates. In formulating a strong dualistic hypothesis we build on the knowledge given below:

1. We can assume that the experiences of the self-conscious mind have a relationship with neural events in the liaison brain, this being a relationship of interaction giving a degree of correspondence, but not an identity.

2. There is a *unitary character* about the experiences of the self-conscious mind despite the unimaginable complexity and diversity of the brain events. There is concentration first on

this, then on that aspect of the cerebral performance at any one instant. This focusing is the phenomenon known as *attention*.

3. There can be a temporal discrepancy between neural events and the experiences of the self-conscious mind. This phenomenon is shown particularly clearly with the experiments of Libet (8) on the conscious perception of a sharp skin stimulus and the interference induced by direct electrical stimulation of the appropriate cutaneous area of the cerebral cortex. The phenomena of backward masking and of antedating indicate that the conscious perception may occur *before* the related neural events. A temporal discrepancy is also revealed by the slowing down of experienced time in acute emergencies.

4. There is the continual experience that the self-conscious mind can *effectively* act on the brain events. This is most overtly seen in voluntary action, but throughout our waking life we are deliberately evoking brain events when we try to recall a memory or to recapture a word or phrase or to express a thought or to establish a new memory.

We also have to build up a philosophy that recognizes the openness of World 1 to influences from the world of conscious experience, World 2. Popper (9) expresses very well the necessity for some loophole in the apparent closedness of the world of matter and energy — that is, of World 1. It is not enough to have an indeterminacy provided by the probabilistic operation at the quantal level. He says that a closed indeterministic World 1.

would be a world ruled by chance. This indeterminism is *necessary but insufficient* to allow for human freedom and especially for creativity. What we really need is the thesis that *World 1 is incomplete*; that it can be influenced by World 2; that it can interact with World 2; or that it is causally *open* towards World 2, and hence, further, towards World 3. We thus come back to our central point: we must demand that World 1 is not self-contained or 'closed', but open towards World 2.

A brief initial outline of the hypothesis may be stated as follows. It is extracted from Popper and Eccles (3).

The self-conscious mind is actively engaged in reading out from the multitude of active centers at the highest level of brain activity, namely, the liaison areas of the dominant cerebral hemisphere. Displayed or portrayed before it from instant to instant is the whole of the complex neural processes, and according to attention and choice and interest or drive, it can select from this ensemble of performances in the liaison brain, searching now this, now that and blending together the results of readouts of many different areas in the liaison brain. In this way the self-conscious mind achieves a unity of experience. This hypothesis gives a prime role to the action of the self-conscious mind, an action of choice and searching and discovering and integrating. The neural machinery of the liaison brain is the World 1 component of the interface with World 2. Furthermore, the self-conscious mind acts on these neural centers modifying the dynamic spatiotemporal patterns of the neural events. Thus we propose that the self-conscious mind exercises a superior interpretative and controlling role upon the neural events.

A key component of the hypothesis is that the unity of conscious experience is provided by the self-conscious mind and not by the neural machinery of the liaison areas of the cerebral hemisphere. Hitherto it has been impossible to develop any neurophysiological theory that explains how a diversity of brain events comes to be synthesized so that there is a unified conscious experience of a global or gestalt character. The brain events remain disparate, being essentially the individual actions of countless neurons that are built into complex circuits and so participate in the spatiotemporal patterns of activity. This is the case even for the most specialized neurons thus far detected, the visual feature detection neurons of the inferotemporal lobe of primates. Our present hypothesis regards the neuronal machinery as a multiplex of radiating and receiving structures, but it does not itself provide the ultimate synthesis. *The experienced unity comes not from a neurophysiological synthesis, but from the proposed integrating character of the self-conscious mind.* We conjecture that in the first place the self-conscious mind is developed in order to give this unity of the self in all of its conscious experiences and actions.

It is proposed that the self-conscious mind plays through the whole liaison brain in the selective and unifying manner. An analogy is provided by a searchlight in the manner that has been suggested by Jung (10). Perhaps a better analogy would be some multiple scanning and probing device that both reads out and selects from the immense and diverse patterns of activity in the liaison brain and integrates these selected components, thus organizing them into the unity of conscious experience. From moment to moment it is selecting according to its interest, the phenomenon of attention, and is itself integrating from all this diversity to give the unified conscious experience. Available for this readout, if we may call it so, is the whole range of performance of those areas of the dominant hemisphere that have linguistic and ideational performance. Collectively we call them *liaison areas*. It has even been conjectured (3) that, when the corpus callosum is intact, there also may be liaison areas in the minor (right) hemisphere.

It might be claimed that this hypothesis is just an elaborated version of parallelism — a kind of selective parallelism. However, that would be a mistake. It differs radically in that the selectional and integrational functions are conjectured to be attributes of the self-conscious mind, which is thus given an active and dominant role. This is in complete contrast with the passivity and ineffectiveness of the conscious mind that is implied in parallelism (11). Furthermore, the active role of the self-conscious mind is extended in our hypothesis to effect changes in the neuronal events. Thus not only does it read out selectively from the on-going activities of the neuronal machinery, but it also modifies these activities. For example, when following up a line of thought or trying to recapture a memory, it is proposed that the self-conscious mind is actively engaged in searching and probing through specially selected zones of the neural machinery and so is able to deflect and mold the dynamic patterned activities in accord with its desire or interest. A special aspect of this intervention of the self-conscious mind on the operations of the neural machinery is exhibited in its ability to bring about movements in accord with some voluntary desired action, what we may call a motor command.

In very carefully designed and controlled experiments Kornhuber (12) has shown that when elementally simple movements are carried out at will without any triggering stimulus, a slowly developing negative potential occurs over the upper part of the brain on both sides, beginning about 0.8 sec before the onset of the movement. It is called the *readiness potential*. In the light of the hypothesis, it can now be proposed that when willing brings about a movement, there is continuous action of the self-conscious mind on a neuronal field of great extent. As a consequence of this action, there is an increase in neuronal activity over this wide area of the cerebral cortex, and then a long and complex molding process leading to the eventual homing in on the motor pyramidal cells that are appropriate for bringing about the desired movement. The self-conscious mind does not effect a direct action on these motor pyramidal cells. Instead, it works remotely and slowly over a wide range of the cortex so that there is the time delay for the surprisingly long duration of 0.8 sec.

The readiness potential indicates that the sequential activity of many hundreds of neurons is involved in the long incubation time of the self-conscious mind in eventually evoking discharges from the motorpyramidal cells. Presumably, this time is employed in building up the requisite spatiotemporal patterns in many millions of neurons in the cerebral cortex, which is a sign that the action of the self-conscious mind on the brain is not of demanding strength. We may regard it as being more tentative and subtle. The self-conscious mind requires time to build up patterns of activity, and it may modify them as they develop.

We can now ask the question: What neural events are in liaison with the self-conscious mind for both giving and receiving? The question concerns the World 1 side of the interface between World 1 and World 2. What properties are likely to give World 1 the requisite openness toward World 2? In our present understanding of the mode of operation of neural machinery anywhere in the brain we emphasize that ensembles of neurons (many hundreds) have to act in some collusive patterned array (13). Only in such assemblages can there be reliability and effectiveness. The isolated responses of single neurons are lost in the background noise. In the cerebral cortex both anatomical and physiological studies indicate that the *columns or modules* provide the operative units (14). As indicated by their name, they have the configuration of a narrow column extending vertically across the 3-mm or so thickness of the cortex. Each module appears to have to some degree a collective life of its own, with perhaps 10,000 neurons of diverse types and with a functional arrangement of feed-forward and feedback excitation and inhibition. As yet we have little knowledge of the inner dynamic life of a module, but we may conjecture that with its complexly organized and intensely active properties, it could be a component of the physical world (World 1) that is open to the self-conscious mind (World 2) for both receiving from and giving to. We can further propose that not all modules in the cerebral cortex have this transcendent property of being modules "open" to World 2, and thus being the World 1 components of the interface. By definition, this property would be restricted to the modules of the liaison brain, and then only when they are in the correct level of activity. For example, all modules would be more or less closed in the depressed brain function of a drowsy state and would be completely closed in deep sleep with a partial opening in dreams. In deep anesthesia and in coma, all would be closed. Also in the extremely high driven activity of convulsions, there is unconsciousness. Thus the modules are closed when there are extremes in their dynamic states, too little or too much.

The modules of the liaison brain could be very specially designed structures, at least in certain areas of the cortex, especially in the linguistic areas whereby the physical world, World 1, achieves an openness to the world of mind, to World 2. I think this is implicit in our hypothesis. There must be some special neuronal structure and action that allows this liaison to occur with operation both ways. If we were to stretch the analogy very far, we could liken the module to a radio transmitter-receiver in such a way that it functions not only for transmitting to the mind, the self-conscious mind, but also for receiving from it. I think this concept is valuable, because we have to stress that the action is both ways all the time. The self-conscious mind is not just passively receiving, it is actively working. In receiving it is active. When it receives, it achieves more action in controlling the performance of the neuronal machinery.

In conclusion, we can return to the argument that the brain-mind problem must be classified as an external frontier of science. Many aspects of this immense problem lie beyond the presently understood principles of science, but nevertheless can be envisaged as eventually coming within the scope of an enlarged science. However, I would agree with the statements quoted from Weisskopf and Bronowski to the effect that there are modes of understanding other than science that give fundamental insight into the human situation and experience — poetic aesthetic, ethical, personal, religious.

I now give some illustrations of problems across an external frontier of science that arise out of my brief review.

1.   An entity, the self-conscious mind, not in the matter-energy world (World 1), surveys the immense on-going complexity of events in the neuronal machinery of the brain (World 1

events). It selects according to its interest and attention and integrates this selection from moment to moment so that there is a unity in the conscious experience derived therefrom. Scientific studies of sensory perception are now at an advanced level, but are handicapped by the lack of a comprehensive theory encompassing both neural events and conscious experiences. In some manner beyond understanding, some neural activities give experiences of light with color, others give sound with harmony, others smell, others pain, others hunger. Some expériences have a correspondence with external events in the matter-energy world, as with light, color, sound, smell. Much is known about the encoding and transmission from sensory organs to the sensory areas of the brain, but *we are completely unable to explain the specific nature of the experiences.*

2. It is an ever-present fact of experience that the self-conscious mind effectively acts on the brain when we plan and carry out actions (voluntary movements) or when we try to recover a memory or to solve a problem. Thus a World 2 entity must be able to bring about changes in the performances of World 1 structures in the cerebral cortex. The influences are very weak and are effective only on the extremely delicate and immensely complicated neuronal operations in special areas of the awake brain. Nevertheless, there is a violation of the strict closedness of the matter-energy world (World 1) that is postulated in the first law of thermodynamics.

3. There must be a partial independence of the self-conscious mind from the brain events with which it interacts. For example, if a decision is to be freely made it must be initiated in the self-conscious mind and then communicated to the brain for executive action. This sequence is even more necessary in the exercise of creative imagination, where flashes of insight become expressions by triggering appropriate brain actions.

4. As mentioned briefly, consciously experienced time may not be in strict correspondence with the related brain events. For example, in listening to familiar music we have the experi-

ence of a blending together from moment to moment of the immediately heard notes with those previously heard and those anticipated.

5.   In deliberate efforts to recover a memory we experience the probing of the "data banks" of the brain and the judgement of the correctness of the memory recall. There seems to be a recognition memory in the mind overseeing the retrieval from storage memory in the brain. Is this another fundamental aspect of the brain-mind problem?

6.   The ultimate problem relates to the origin of the self, how each of us as a self-conscious being comes to exist as a unique self associated with a brain. This is the mystery of personal existence. We come to exist as a self-conscious being because of immersion from babyhood in the cultural environment (World 3) in which linguistic communication is preeminent. The brain (World 1) plus the cultural environment (World 3) are necessary for the development of the conscious self in World 2, but I have argued elsewhere (15) that the uniqueness each of us experiences can be sufficiently explained only by recourse to some supernatural origin as well. Coming-to-be and ceasing-to-be are linked components at the ultimate level of the brain-mind problem. What happens to the conscious self at brain death? Its wonderful instrument disintegrates and is no longer sensitive to its cognitive caresses. Is the self renewed in some other guise and existence? This is a problem beyond science, and scientists should refrain from giving definitive negative answers.

I would like to close with a vision of the future expressed in an article, "Quo Vadis," by Bridgman (16), a distinguished Nobel Laureate in Physics:

> It seems to me that the human race stands on the brink of a major breakthrough. We have advanced to the point where we can put our hand on the hem of the curtain that separates us from an understanding of the nature of our minds. Is it conceivable that we will withdraw our hand and turn back through discouragement and lack of vision?

REFERENCES

1. V. F. Weisskopf, The frontiers and limits of science. *Bulletin of the American Academy of Arts and Science*, **28:** 15–26, 1975.

2. E. P. Polten, *A. Critique of the Psycho-Physical Identity Theory.* Mouton Publishers, 1973.

3. K. R. Popper and J. C. Eccles, *The Self and Its Brain.* In press, 1976.

4. K. R. Popper, *Objective Knowledge: An Evolutionary Approach.* Clarendon Press, 1972.

5. J. Bronowski, Science, Poetry and "Human Specificity" (An interview by G. Derfer). *The American Scholar*, **43:** 386–404, 1974A.

6. J. A. Bronowski, A Twentieth Century Image of Man. *Leonardo*, **7:** 117–121, 1974b.

7. E. P. Wigner, Two kinds of reality. *The Monist*, **48:** 248–264, 1964.

8. B. Libet, Electrical stimulation of cortex in human subjects, and conscious memory aspects. In: *Handbook of Sensory Physiology*, Vol. II, edited by A. Iggo, Springer-Verlag, 1973, pp. 743–790.

9. K. R. Popper, Indeterminism is not enough. *Encounter*, **40:** 20–26, 1973.

10. R. Jung, in *Brain Mechanisms and Consciousness*, edited by J. F. Delafresnaye, First C.I.O.M.S. Conference, Blackwells Scientific Publications, 1954.

11. H. Feigl, *The "Mental" and the "Physical,"* University of Minnesota Press, 1967.

12. H. H. Kornhuber, Cerebral cortex, cerebellum and basal ganglia: An introduction to their motor functions, in *The Neurosciences: Third Study Program*, edited by F. O. Schmitt and F. G. Worden, MIT Press, pp. 267–280, 1973.

13. J. C. Eccles, Functional significance of arrangement of neurones in cell assemblies. *Arch. Psychiat. Nervenkr.*, **215:** 92–106, 1971.

14. J. Szentagothai, The "module-concept" in cerebral cortex architecture. *Brain Research*, **95:** 475–496, 1975.

15. J. C. Eccles, *Facing Reality*, Springer-Verlag, 1970.

16. P. W. Bridgman, Quo Vadis. *Daedalus*, 85–93, in *Proc. Am. Acad. Arts and Sci.*, **87,** 1958.

DISCUSSION

General Comments

Dr. Edelman. I would first like to say how much I admired the efforts of Sir John Eccles. His comments were extraordinarily stimulating and refreshing, and he did a service in bringing back to our attention this ancient and pervasive problem. I think that even though my remarks are going to strongly disagree with his interpretation, it's not for lack of admiration of where his emphasis was placed.

The first thing I would like to say is that while we do not know what the *mind* is, it cannot be dismissed as a nonproblem. There is obviously a problem, and it's this that I admire about Dr. Eccles' attempt to say that there's no sense putting it aside. It is one of the most important problems we face in or out of science.

The second thing I would like to say, however, is critical. To describe our ignorance in terms of some other category of ignorance, namely a self-conscious mind, seems to me to only restate our ignorance. It doesn't seem to me to solve the problem to divide the world as Popper has into Worlds 1, 2, and 3. I believe there is one world within which a set of events occurs and within which these entities that we're discussing exist.

The main point, however, that I'd like to make is that whatever the mind is, whatever we attribute to what Dr. Eccles calls the self-conscious mind, it is an ensemble property. That is to say, it is a property of a collection of things, a very large collection of things, just as, for example, temperature is an

---

Editor's Note. Asterisks denote a break in sequence or a change from one discussion panel to another.

ensemble property of molecules. Temperature has no meaning if you reduce the ensemble to a single molecule. I think we do know operationally that's true of the mind, and I think Dr. Eccles would agree with that.

The second thing I'd like to say about the mind is that it also seems to me to be a structure dependent property. It's uniquely attributable to a certain structure. I don't think Sir John would disagree with this, either. Here I think the problem is rather complicated, because the activities of the structure are expressed in terms of a code. There is a neural code, and I was reminded when I listened to Dr. Eccles' remarks of the ancient biological controversies involving the genetic code. There has always been a long-standing argument about preformation and epigenesis. Namely, in the spermatozoan there was supposed to be a little man, inside which was a spermatozoan, inside which there was another little man, and so on and so forth. This was before we knew that, in fact, the genetic code was what was responsible for the expression through a very complex machinery of protein molecules to make our bodies. I believe that in the same way the transduction of what amounts to a physical image in one's eye must be in terms of some code, and that this code must be uniquely structure dependent. The challenge that I would put to Jack Eccles is that if he agrees that it's structure dependent, does he really insist that it is also composition dependent?

I, for one, believe that if you knew the structure of the brain and you represented that structure in some other materials capable of the same functional events, what you would get is what he calls the self-conscious mind. And here I want to use an example from my own field. We all have molecules in our bodies and a system known as the immune system, which is capable of recognizing foreignness; that is, it is capable of distinguishing between self and nonself.

Prior to a certain date, the system was believed to work by instruction. That is to say, a foreign molecule which had a certain shape, for example, a protein molecule on a virus, would enter our body and impress its shape on the antibody molecule of protein inside our body much as a cookie cutter stamps its shape into dough. This shape was then said to be instructed, and the next time around it would recognize the foreign molecule.

Well, it turns out as a result of the elaboration of the structure of the immune system, that this isn't the way it works at all. The way it works is that you already have all of the information you're going to need to recognize any structure in the world before you ever see those structures. Now, that sounds quite silly, but in fact the way it works is this. You have about 10^{11} cells in your body, each one of which carries a molecule of a certain different shape, much as a lock fits a key. When a foreign object comes from the outside that you've never seen before in all your individual life or in the evolution of the species, it scans the system and fits some of the shapes. When that happens, a selection takes place, and more of those cells are made. They divide and make daughter cells.

Now, this system has the properties of memory; it has the property of recall; it has the capacity to distinguish positively by naming, if you will, one thing or the other and recalling that recognition. But, it is not a cognitive system, and if you did not know its structure, you would have posited a great number of wrong instinctive theories about how it works.

We now, in principle, understand how it works, and I would suggest perhaps that an analogy be made with the brain. That is to say, the antibody system which can do the same recognition in very many different ways operates in such a way as to convince us that maybe instruction took place when it didn't. I only suggest this for your thoughts when you think about the brain. It may not be a machine which has a unique answer. Like the minimal system, it may be a selective machine which is degenerate and carries out thinking in very many different ways.

The other point I would like to make about the remarks that Dr. Eccles put before us is this. If the only instrumentality for examining the nature of the self-conscious mind is the brain itself, then I'm afraid it already lies outside the sphere of operational science. I'm afraid we can't really do much operation on it. I say this in the light of his last remarks about the self and whether it lies beyond science or not. I feel the issue is whether he has posed before us an hypothesis that can be tested. In my opinion, I do not think so, although he has posed the problem in very lucid and exciting terms.

* * *

Dr. Cooper. I think Professor Eccles stated some of the difficulties of understanding the central nervous system in a concise and very brilliant way, and I find myself agreeing with his statement of the difficulties. But I disagree entirely with his conclusion that one has to introduce a World 2 in order to understand the brain or the mind.

Dr. Eccles began by saying that one had to reject what he called parallelism. What he means by parallelism, as I understand it, is that one attempts to construct even the conscious mind out of ordinary materials. Now, of course, that is a very difficult enterprise. As Dr. Eccles said, "When you have something like a thousand million neurons all firing, how can one understand something so complex?" Well, I'm not denying the complexity, but in fact we have understood things where the numbers were as large. For example, as physicists we have understood the behavior of gases, and in gases there are 10^{23} molecules. This is a very large number. Yet, the behavior of these molecules can be understood. Now, this doesn't mean that we are sure that we will be able to understand how the brain functions, but rather that understanding is not denied to us in principle by the ordinary means of science.

Related to this Dr. Eccles asked a very amusing question. Can a brain understand the brain? That sounds like a paradox, but the answer is, obviously, yes. Why do I say that? If, in order to understand the brain, we have to keep a billion, billion things in our head at the same time, then we may not be able to do it, because we don't have the capacity. But if the brain operates according to relatively few principles, then it's perfectly possible for our own brain to understand these few principles. This has been done over and over again in other aspects of nature. Again I refer you to the behavior of a gas. If one wanted to follow the individual paths of all of the molecules of the gas, no one could keep that in his head. But understanding a gas is not equivalent to following all the molecules. We can understand all kinds of things about the gas with a few relatively simple principles.

Dr. Eccles said that if we use the usual methods of science, then our behavior would be determined, and he rejected determinism as a possibility. But, there is nothing in our experience that is inconsistent with a complete determinism. What we have to explain is not necessarily freedom of will, but our sense that we have freedom of will. As Spinozoa said (remember this is 200 years before Freud), men believe that they are free because they are conscious of their actions and unconscious of the causes whereby these actions are produced.

Professor Eccles also said in rejecting reductionism or parallelism that if you took this approach, then it would be very difficult to understand such things as human creativity. This is not true. Creativity is completely consistent with complete determinism; they are logically independent concepts. Consider the following: everyone would agree that Mozart was creative. That to me is the definition of creativity. There's nothing that says that God could not, when He created the universe, have known that Mozart would live and that Mozart would compose his 41 symphonies and so on. I'm not saying that's the way it happened, but they are logically independent concepts. So, there is nothing in determinism that precludes human creativity, or human imagination for that matter. Again, what I'm really saying is that we cannot reject the possibility of the type of explanation which comes from World 1 to explain these phenomena.

I'd like to conclude with something people often say to me when I talk like this; they feel disappointed. They say, "Suppose that you're right. Suppose you really could explain what a human being is, and explain consciousness and self-consciousness using ordinary materials like neurons. Doesn't that make us something less? We'd be only neurons or only chemicals." I reject the use of the word "only." It reminds me of the people that used to say with a sneer that you are only made of 92 elements, which cost about $1.02. That to me is equivalent to saying that a Shakespeare play is only words, or a Mozart symphony is only notes, or the cathedral at Notre Dame is only stone. Yes, in that sense they are made of perfectly ordinary materials. What else is there? Well, what is there is clear. It is the organization of the words, the organization of the notes, the marvelous organization of the cathedral. In effect, in what we are proposing we don't necessarily reduce ourselves at all. If, in fact, it turns out that we can understand even ourselves by ordinary scientific means, and if we are only made of ordinary materials, remember that these ordinary materials are organized in a manner that has to be regarded as extraordinary.

* * *

Dr. Bethe.   Let me begin by saying that the idea that there may be a breakdown in these laws of physics in the relationships between World 1 and World 2 is totally unacceptable to me, and I think to most natural scientists. I think it's quite unnecessary for the things Dr. Eccles wants to explain. I want to mention in this connection other recent developments in the biological sciences which I know only very peripherally, since I'm a physicist. In the early nineteenth century people believed that organic substances, substances which contained carbon, were intrinsically different from inorganic substances which don't contain carbon. This was very soon shown to be incorrect. You can synthesize all sorts of organic substances. Then only a very short time ago there was a tremendous controversy whether living things obey different laws of physical chemistry from inanimate objects. I think it has been shown conclusively that the same laws hold. The same physical and chemical processes occur in life as they do elsewhere. So, I am firmly convinced that this process will go on, and that the validity of the laws of physics and chemistry will extend further to the function of the brain and to the things which we call the mind. In fact, in Dr. Eccles' talk there was very much to encourage me that research is already advancing quite far in that direction.

In addition, Dr. Eccles emphasized that the World 1 which we perceive, the world of phenomena around us, appears to be different from the world of the mind which we directly experience. This again is something which has an analog in physics. This analog is complementarity. In the world of physics, as probably most of you are aware, it is now thought that every piece of matter is both a particle and a wave. It has been possible to find the underlying laws which tell us when particles are just particles moving according to Newton's laws, and when they act as waves such as light waves.

There are many other things in the world which can be looked at from two entirely different points of view. The point of view of World 1 and World 2 look entirely different, and yet they describe the same entity. Two aspects of the same entity may be complementary, and only by understanding both aspects will we completely understand the phenomenon. So, in my opinion, we can draw on enough known experience in biology and known experience in physics and chemistry to encourage us that there is a solution in sight; that there is a solution somewhere in the distant future which will reduce everything to one monistic world.

*   *   *

Dr. Schwinger.   I would like to make some brief remarks about some aspects of Dr. Eccles' talk which struck me particularly since they were aspects of physics. Dr. Eccles referred to a quotation by Popper, which as I understood it essentially dismissed the idea of determinism as a tautology. Determinism or causality, the two words are roughly mixed together, and I'm not sure which one we're really talking about, but I would just like to point out the fact that in the history of physics the understanding of causality has actually changed. Causality, as it was understood in classical physics, was the notion that if you knew what all the particles were doing in a certain system and how they were moving at any particular instant, you could then predict with total precision where they would be and how they would be moving at any later time, or indeed at any earlier time.

But we have now known for almost 50 years that this is simply not true when we consider the atomic world. The concept of causality has changed. I'm referring, of course, to the uncertainty principle: the limitations of knowledge which are not due to our stupidity or ineptness, but are inherent in the very foundations of nature.

Despite that limitation, there still remains a concept of causality, but one which is much more sophisticated. It is not a statement about where the particles are, and how they are moving, but rather refers to a wave function, a very remote mathematical concept. But it is still true that if you know the wave function at one time you can predict it at another time, and certain physical and objective facts follow from that. In other words, if the significance of causality has actually changed, then it's hardly a tautology. It has physical, objective meaning.

Dr. Dean.   In the history of theology, there has been a long-running battle from which theology has been steadily retreating. It's now referred to as the theory of the "God of the gaps" — you invoke God if you have gaps in knowledge. In other words, if we don't understand something now, must we be so committed to immediate explanation that we'll seek an explanation by invoking the supernatural, or will we withhold that kind of conclusion and keep looking for a natural explanation?

I'm wondering if that question would apply not only to the mention of the supernatural in the last page of the printed lecture, but also to the whole argument of moving from World 1 to World 2. Is there really a point at which the complexity of the neurological phenomena seems so great that we <u>must</u> assume that there is a self-conscious mind directing the functions of the brain? I am reminded particularly of when Dr. Eccles was talking about the unitary nature of consciousness, and of how difficult it is to explain the unitary nature of consciousness by reference to the brain alone.

Dr. Eccles. I've been wanting to comment for some time. I don't like standing up here and getting all these spears thrown into my body without resisting. First, about the supernatural, my point is simply this. I am an evolutionist. I believe in evolution and in genetics. I believe the evolutionary process was the most marvelous biological mechanism for eventually building the genetic code that makes human beings, including myself. That is, I have a long biological ancestry, you might say, that builds my body and my brain. No problem there, there is nothing supernatural required in that story. But what about the inner experience that I have? I can explain all of you entirely on natural grounds if I wish to, but, as we communicate, I would discover that you are conscious beings like I am. It's this inner unique experience as a conscious self that I'm talking about. Now you might say that's just the same thing that would result from building a brain. The brain would become conscious, and that would be you. If it had been another brain, it would have been somebody else.

This, I think, is quite inadequate. First of all, the genetic code that builds the brain of an identical twin is the same, but they are different selves. They look alike, but to themselves they are completely different and distinguishable. So, the genetic code is not adequate for the distinction and the uniqueness of these two selves.

Well, you could say, let's look at it the other way. You are unique because you have a unique genetic code. You have a DNA structure that built your brain, and another DNA structure would have built another brain which would have been somebody else. So, I am unique because I *have* a unique genetic code. That is granted, but do you know how unique your genetic code is? It's beyond all imagination. That is, this explanation becomes absurd when you realize that a conservative estimate is that there are 10 to the 10,000th possible genetic codes to build human brains. This is, of course, a mystery, and that is why no one likes it. But we are a mystery. We are infinitely mysterious — each human being is. The idea that we just have some factual explanation of our origin and our nature and our destiny is something which I think is too naive for us to accept.

I agree that when you're down at the level of a fundamental particle, determinism and causality become statistical operations. But what I want to point out is that, when we're talking about the specific operations of the human brain, we're talking about enormous assemblages of molecules in a patterned array, not fundamental particles. The mass is so large that it comes beyond anything at the uncertainty level. I once did sums to try to see how far the Heisenberg principle of uncertainty would hold in giving some degree of variance for the operation of a single synapse. Again, it misses by many orders of magnitude. This tells you that you're in order of sizes for which the uncertainty at the particulate level is, I think, too small to matter. I think we are up against quite another level of the problem when we come to the way in which the World 1 is open to the influences of World 2. The masses are too large to be involved in that, and the uncertainty principle doesn't help us, because this kind of determinism, again, gives you no foothold for a willed operation.

Ðr. Dean, the question of the influence of World 2 upon World 1 is not supernatural in the sense I would use it. I think we are just making a larger expansion of what we call natural. I think World 2 is investigable by scientific methods and belongs to the natural order because it is our experiences we're talking about, and we are in nature. It just happens to be different from the world of matter and energy.

It was terribly important in the early days of science to have "something taken off." You had to simplify your program so that you could apply rigorous testing and investigation. But, I think we now have to realize that we are officially rejecting an important part of nature for this simplification, and now my effort is to enlarge science so that it can begin to come to terms with this additional aspect of nature, namely, the world of conscious experience.

Science and Philosophy

Dr. Cournand. It seems to me that the challenge that Dr. Eccles has put forward is whether there is a possibility of reconciliation between belief, on one hand, and understanding of the world and of one's self by means of science, on the other. As far as I know, this is the first attempt to propose an experimental approach to this problem.

Up to now, scientists have completely separated knowledge from religion. Pasteur himself, who was a great scientist, was also a man of belief, but he never tried to mix them together. Aware of Dr. Eccles' past scientific achievements, we may feel assured that he may contribute important information in his attempt to reconcile two apparently unrelated approaches to the search for a unique truth. Incidentally, I would like to suggest that the notion of self-consciousness represents a model of the living form of an immortal soul. This means a revival and confrontation of the separate entities of matter and of soul.

Dr. Harvey. I think Dr. Eccles has recreated a kind of classical dualism that really is not very productive philosophically, and I myself would think, though I'll defer to my scientific colleagues here, is also scientifically unproductive, because it doesn't give you any way of testing or checking it out in terms that scientists recognize. He postulated, for example, a self-conscious mind which somehow floats above the brain and scans it, but yet remains in contact with the brain. Further, it is assigned a supernatural origin at the end. But, you also have to account for someone to put this self-conscious mind in contact with these brain states. Now, it might sound funny for a so-called theologian to say this, but I'm simply saying that this is a candidate for scientific work. I think theologians or philosophers ought not to get into the business of being driven to supernatural explanations in order to account for scientific phenomena.

* * *

Dr. Barbour. I think we are in agreement on rejecting much of the dualism that Dr. Eccles was outlining, but I would want to insist that these two languages will be with us indefinitely and that the two aspects are not just a temporary feature of our present ignorance. I would actually interpret Bohr's complementarity principle as saying this same thing. Because of conceptual limitations in our ways of thinking derived mostly from the everyday world, when one gets down to the subatomic world, this duality, even though it's unified at the mathematical level in quantum mechanics, is going to be with us. So, this sort of monistic single system will perhaps only relate the two worlds.

There were two or three references to physical and chemical laws explaining everything eventually. I don't think they will. I think what happens is that when you relate two levels you have to have a new conceptual structure that isn't present in either of

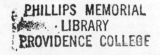

the two levels that you are relating. It isn't a matter of reducing the higher level to the lower as in the present scheme. When one says that one will explain everything in terms of physics and chemistry, I'm dubious.

* * *

Dr. Schwarz. I think when you talk about what makes me want what I want, the underlying question is one of freedom and necessity, in the same way that you have body and soul, or as Dr. Eccles talked about it, mind and soul. I have in my limited layman's knowledge of science a couple of other "polarities" which are used to describe the results of scientific investigation, the duality of light, for instance, the duality of matter to some extent, and what I see here is perhaps a "both-and." On the one hand, there's freedom; on the other hand, there's necessity, and as I read this paper through it seems to express this kind of "polarity." He says, no, it's not that simply nothing is determined, but it's not just all determined, either. There is something we can't get our hands on, because we don't have the conceptuality.

This is what one of us mentioned here. It is a conceptual tool he tries to propagate here, to divide into three worlds, hoping that then the phenomena will become more intelligible.

We are confronted with problems for which we don't yet have the conceptual tools. Perhaps somebody will find them, and then we will see that both ways of getting at them fit. But to absolutize one of these, and this is what he warned about, and say absolutely no, would pervert the whole thing. I think this kind of openness must exist.

* * *

Dr. de Duve. As a scientist and a life scientist, I would say that he has approached the problem from the philosophical point of view and not from the scientific point of view. In other words, he has given a philosophical analysis of a problem, and since I'm not a philosopher, I don't think I can really criticize him on philosophical grounds.

I would say, however, that if I were involved in the kind of research that Dr. Eccles has been talking about, I would not take as a working hypothesis his hypothesis of two worlds: his dualistic hypothesis which implies one world, the material world which is accessible to our instruments and to our scientific investigation, and another world which, by definition is not accessible to these investigations but which somehow interacts in a mysterious manner with the first world in certain specialized areas of the left hemisphere of the human brain. I'm not saying the hypothesis is right or wrong. I say I would not adopt it as a working hypothesis because it is a sterile hypothesis from the point of view of the scientist. It cannot be tested experimentally.

Dr. Cobb. I do think that Dr. Eccles' paper is a very philosophical one. But, I'm a little troubled if we separate philosophy and science quite so radically that we suppose that when a scientist is a materialist there is no philosophical assumption involved, but that when he is a dualist there is a philosophy. It seems to me there are philosophical assumptions that have undergirded the models and the methods of the natural sciences, and very successfully so. I'm not challenging that at all. But at certain frontier points, questions arise, as they have arisen in the past, with respect to the philosophical undergirdings of particular forms of science which lead persons investigating certain fields to want to experiment with models that are derived from other philosophical traditions. In that sense, I don't think we should simply say this is a philosophical paper, and obviously not scientific, and therefore irrelevant to the further development of science. Certainly, I don't think Dr. Eccles intended it that way, and I believe he is able to explain what he is doing in another way.

There are other physiological psychologists, and the only one I know personally and have had some discussions with is Roger Sperry. He's interested in many of the same phenomena and explicitly thinks that one must take account of conscious experience as a causal factor in what occurs in the brain, which is, of course, excluded by traditional materialism. He would, however, reject the dualism which Dr. Eccles affirms. He prefers to use a whole-part model. He speaks of the conscious experience as a wholistic phenomenon and then insists that you cannot understand the whole simply in terms of the behaviors of the parts.

Now again, I'm being very simplistic, but that's one other kind of a philosophical model that he finds he needs to employ in order to devise further experimentation in the expansion of the science in which he's working. He refuses to say that's mere philosophy. In fact, he is inclined to say that's not philosophy at all. This is pure science. I'm always skeptical when people are doing pure science, and I see something there that looks to me like it belongs to some of the philosophical traditions. I hope that we will not be so bound in the identification of science with any particular metaphysical framework of thought that we assume a priori it would not be science, if a different conceptuality were involved.

I think in the history of science we do see new paradigms emerging which sometimes involve quite radical reshaping of what seems to be scientific. If Dr. Eccles is right or Dr. Sperry is right, then the change in the total understanding of science will be very considerable.

*   *   *

The Future of Science

by

LANGDON GILKEY

Our topic, "The Future of Science," can be interpreted in at least two ways. It can mean the future *content* of science, where it is going, what scientists will discover in the future, the new understanding of the world it will provide. On this subject I have nothing at all to offer. The other meaning concerns the role of science in the wider cultural context, as a factor in the social life of man, as a force in future history. This, too, can be an important and interesting subject, one, possibly, on which a nonscientist can offer some relevant thoughts. The reason is that science is not only a method of inquiry pursued by scientists and a consequent body of tested hypotheses. It is also an historical force of overwhelming significance, shaping the social existence of mankind in ever-new directions. It has transformed not only the character of the lives of men and women, but also their understanding of themselves, their understanding of the history of which they are a part, and thus their view of their destiny. Strangely and unexpectedly, it has, moreover, had a career in modern life not unlike that in former times of my own discipline, religion. Our question, then, concerns the role of science in our common social future.

The creative effects of science are spread before us in every aspect of modern social existence. It is through science that technology — which is itself as old as Homo Faber — has developed to its present, astounding levels; and it is through that technological development that industrialism has transformed every facet of our lives, personal and social. Many scientists resist this close identification of pure science with technology and emphasize the distinction between the disinterested search for theory and the practical application of theory to concrete problems of everyday life. Such a sharp dichotomy between scientific understanding and technological practice is, however, untenable. Both the history of modern science and a theoretical interpretation of it show that this dichotomy, generated out of the Greek understanding of knowing, is inapplicable to modern forms of cognition. It is as unconvincing as is the argument of the theologian that his theological understanding is unrelated to organized religion, to the Church, and thus his profession is in no way responsible for the historical excesses of religion.

Few funds for pure science are granted without an eye to practical application, as every lobbyist in Washington knows; nor would major research be *publicly* funded, were it not for its technological possibilities. And no discovery in the most esoteric branch of pure science is announced to the public without specification by the discoverers of the potential "revolutionary practical uses" of the discovery (1). On the theoretical level, the great John Dewey was surely right that the essence of the cognitive method of modern science, as opposed to that of the Greeks, was to unite manipulation of the perceived environment with theoretical understanding so as to achieve control over a puzzling situation, a control to be achieved through knowledge, tested by application, of the causes and factors at work there. To know, and to be sure that you know, is to be able to control through intelligent foresight and so understand the actual course of events. Thus, he argued, science and technology are in essence one and the same operation of organized intelligence on its world (2).

Above all, as Francis Bacon said, empirical knowledge, knowledge of the forces that surround us in our world, means the power to control and manipulate those forces for human well-being. Greater knowledge always means greater power. Thus, whether this be their intention or not, the "knowers" in a society bequeath to their culture ever new powers to transform its life. It is for this reason that knowers, religious or scientific, are valued as well as revered by their society, the priest's robes and the scientist's white coat signifying much the same social role as the knower of significant secrets and thus the doer of all-important deeds. Modern empirical science has been well aware of this; and at each stage of its career, including the present, it has justified the social status and role of science and of the scientist by appealing to the vast potentialities latent in scientific knowledge to remake human existence for the better. Just as theology cannot without internal and external contradiction disavow a concern for human salvation, neither can pure science disavow technological application without denying its own essence as a cognitive discipline and without abdicating its predominant, creative, and funded role in our social existence.

The positive effects of science on modern culture have been, however, by no means purely materialistic; they are more than the results of technology and industrialism on our material standards of living and so on our health, comfort, and general security or well-being — great as these latter are. For in my view science has been the most important formative factor in creating what we may call the modern Geist. By the modern Geist, or spirit, I refer to that view of man, of his world, of his possibilities, his history, and so of his destiny in the future which has distinguished — as have also technology, industrialism, and their effects — modern culture from other cultures. A most crucial attribute of modern science has been its capacity to know what has not been known before, to be, therefore, creative of new knowledge, new understanding, new concepts, new views. A culture dominated by scientific knowledge has, therefore, developed a critical relation to tradition, even its own tradition, and a tolerance of the unaccustomed, the unorthodox, the deviant that is itself something new in history and a most precious aspect of modern culture. But new ideas mean, as we noted, new possibilities for life, new forms of life, a new and remade world, a new future. Thus again out of science has come a new understanding of human possibilities, of the capacity of man to reappraise and remake his world, and of history as the locus of these novel possibilities. The American and the French revolutions in the eighteenth century as the enactment of the *new* in history, and in the nineteenth the theories of historical progress and of dialectical materialism as the march of history toward new and better possibilities are, alike, inconceivable without two centuries of modern science. As the Judeo-Christian understanding of history provided the hither or long-term foundation for modern science, so the latter provided the most important nether base for the Enlightenment and the modern culture that has developed out of the Enlightenment. From this has arisen a new understanding of man as capable of controlling natural forces for his own use, of remaking his social and historical worlds and so of history as a realm of promise. If man in former ages felt himself to be the victim of forces he could neither understand nor control, men and women through scientific understanding and technological control have come to feel themselves the masters of these forces and thus even of their own destiny. Science has given to men and women a consciousness of their own freedom in nature and in history unknown before, and out of this new self-awareness has come the buoyancy and the hope for the future characteristic of modern culture wherever it has penetrated.

The full defense of this thesis would consume all our time. Suffice it to say that science has not only remade our natural and social environments, it has also remade ourselves and our views of the history we live in, giving us a new confidence in our powers to know, to create and remake our life, and a new hope that through our capacity to know we can master the implacable fates and create, at last, a humane world. Moreover, science has represented in our culture a most precious human attribute, the love or eros for the truth and the intrinsic joy — not to say ecstasy — to be experienced in relation to the truth. Thus it has provided our modern culture with much of its spiritual grandeur and has given opportunity for countless persons to fulfill their lives in selfless commitment to the truth and to it alone. Science has shaped a significant and enduring form of human authenticity in our time. As a cultural force, therefore, it has been as creative of our spiritual existence as of our material: of our aims and techniques of education, of our understanding of ourselves as knowers and doers, of our views of morals and of human authenticity, of our confidence in history and in the future. There is no part of our cultural life, including religion, that has not been transformed, and creatively so, by the impact of science. The question of the future of science is, therefore, not only a question of our material future, it is also a question of the future of our modern free, diverse, creative, and confident society in all its many aspects.

Few things in human life, however, are creative in moderation, or — even more important — are moderate about their own creative powers. Because it brought a quite new and more reliable sense of certainty in the knowing process, a new freedom from tradition and from absolute spiritual authorities, and a new confidence in the power of our freedom to control our future, science has appeared to much of our recent civilization — to laymen, educators, philosophers and scientists alike — as *the* salvific force in history. This faith still inspires and supports much of the scientific community. If only, said John Dewey, we could apply the scientific method and spirit to all our problems, those problems would recede; science can save us if only we harken to her (3). It was the two roles of science we noted that gave the new method of science this apparent saving power: as the bearer of testable and thus valid knowledge and as the key to control. As with all saving religions — science has been a *religious* force for our culture as Marxism has been for another culture — sacred knowledge establishes and guarantees the power to control whatever menaces us. Through that sacred knowledge there is given to those who bear it — be they yogian, priests, or scientists — mastery over the fates and so the key to future well-being. In the case of medieval religion, the sacrality of knowledge came from its divine source in revelation and from its power to save us from sin and death. In the case of science it came from the objectivity of its method, the sharability of its conclusions, and its utility in technological application to the pressing problems of everyday life.

Whenever knowledge and control have such a sacral character — that is, whenever they promise salvation from what we take to be our most fundamental ills — they dominate the culture that forms itself around them. As religion had dominated the civilization of the medieval period, so therefore science has dominated ours. It has determined education, molded our sense of human excellence, grounded our hope for the future, and established itself as the queen of all the other disciplines of learning. It became quickly *the* method of inquiry according to which all the other Wissenschaften must remake themselves or be excluded from the academic court. Its empirical and objective techniques represented *the* form of knowledge to which every other mode of knowing had to conform or be banned from serious consideration. It alone, therefore, defined what was real and effective in intelligible and thus rational experience. As logical positivism, the philosophical counselor, advocate, and handmaiden of the new queen said: Existential statements, statements of "what is the case in the world," are scientific in form, or else they are meaningless. Relevant reality is known and dealt with only in *this* way; all deviant claimants to knowledge, be they aesthetic, moral, philosophical, or religious, are merely emotive and thus tell us nothing of what is real. The other disciplines quickly fell in line with the new queen; psychology and the study of man, social and political theory, literary criticism, history, philosophy, and, heaven help us, even theology sought to become "scientific" if they were to be recognized at all in academia — as in the medieval period every discipline claimed to have a theological foundation as the guarantee of its validity and of its usefulness to human welfare. And, as with all queens, there were rewards. As the church ended its reign owning one-third of Europe, so science in the modern university receives and uses the vast majority of private and public funds. Far from seeming irrational or unjust, all this is to us as obviously reasonable as the corresponding role and stature of religion was to medieval man. Is it not science alone that promises valid knowledge if such knowledge be possible, and is it not scientific understanding alone that can guarantee our control over nature and the remaking, if again such be possible, of our psychological and social existence? Sacral knowledge and the power it gives over all that seems to threaten us makes of any discipline a queen. And let us note, just as it was not only the theologians who made theology the queen but people of all sorts who revered as sacred and thus as saving the knowledge formulated in theology, so it was not scientists alone or even their spokesmen who brought science to this position of dominance. It was also those in all walks of life who found scientific method to be the key to valid knowledge in every field and to contain the cherished promise of greater human well-being.

Queens, however, are not always so; they have their day and then decline, possibly into banishment. Coup d'états, palace revolts, and changes of rule take place in cultural as in political life. The development of modern culture since the Reformation and the Enlightenment has seen not only the rise of science to cultural dominance; it has also witnessed the decline of the Church as foundational to social existence in all its aspects, and correspondingly the eclipse of theological understanding as the ground of every valid field of inquiry and thus sovereign over all. As everyone familiar with this history knows, there were innumerable causes of this loss of ecclesiastical and theological sovereignty. Central among them, I believe, was the fact that the Church and its truth claimed an absoluteness that nothing human could or should claim. To be sure, the God to which Christian faith witnesses and the divine grace it proclaims are, I believe, absolute. But the human response to God, and thus the historical forms of Christianity, its doctrines, its moral laws, its human institutions and clerical authorities — these were not at all as absolute, unambiguous, and "pure" as their representatives, in their enthusiasm for the saving power of religion, believed. Thus the claim to absoluteness led in the end only to disaster for religion and for the society in which it was central. Absolute faiths clashed with one another for two centuries; absolute moral laws bred hypocrisy, injustice, and, in the end, irrelevance; absolute spiritual authorities found it necessary to crush every evidence of developing autonomy of mind and of spirit. And thus in the end religion was dethroned. Some concluded from this process that every form of religion was destructive illusion, so that religion could be successfully eradicated from intelligent existence. That this view was wrong has been shown by the fact that religion has reappeared, in even absolutist and destructive forms, in our political, economic, and social existence, and that other queens, for example, science or politics, now preside with equal absoluteness over the academic courts of our modern world. Others slowly realized that human religion, although an essential aspect of our life as creatures of God, is still *human*, influenced by culture and thus characterized by the relativities of history. Thus, so they argued, if she be true to herself, religious faith has no business claiming the right to control the wide varieties of human existence, nor does it have any call to become ruling queen of the diverse and free sciences.

Now the main point of these remarks is that in our day we are, I think, witnessing a similar process: the dethronement of the most recent queen. One could cite many evidences of this shift in cultural sensibility. The rise of the interest in the occult and in mysticism, so astounding among our educated youth, represents a direct challenge to the supremacy of the scientific consciousness and the world view it has created. And the general disillusionment with technology bespeaks a deep questioning of applied science as the answer to human problems. Any visitor to meetings of the NSF can feel there — even in that center of social prestige, economic prower, and political clout — a new nervousness and uncertainty about the role of science, about, in fact, its predominance in our cultural life. The deeper causes of this widespread uneasiness about science lie, I believe, in the same profound "fault" evidenced in the career of the erstwhile queen, religion. There we saw that religion "fell" from sovereignty because, although it is an essential and very creative aspect of human existence, it made itself absolute, predominant over the other aspects of life and the sole source of knowledge and of healing. I believe the same has been true of science, also essential and creative; I shall seek to show this, and its ambiguous consequences, in relation to three important aspects of science as a cultural force.

There is, first of all, the question of the absoluteness of the scientific consciousness as the entrance into what is actual. Is the method of science the *sole* cognitive avenue to the real? Or, put in terms of our court analogy, should science — or any other discipline — be the ruling queen? Although a great deal of valuable philosophical thought has been given to this question since the rise of science — one thinks in our time of Husserl, Whitehead, Buber, and Tillich — I shall argue my point in terms of the career of science itself.

As we have seen, science has represented in our cultural life an intense and continuing experience of human self-transcendence: of the power of human inwardness and autonomy to rise above all prior conditioning to know the truth "objectively," of the power of inward commitment to the truth to transcend tradition and authority in order to achieve new concepts, and of the power of informed intelligence creatively to remake its world. The sense of the self as free, potentially "objective," and thus creative within the stream of events has been uniquely characteristic of our culture, of Enlightenment culture. It is the direct result of science, or, better, of the experience of being a scientist within the scientific community. Modern man has *known* of his own potential freedom over his own prejudices, his own baser desires, over tradition and conditioning, and over his world largely through the experience of themselves as knowers and manipulators which the members of the scientific community have for several generations enjoyed.

Here, however, a strange contradiction enters, one that is most significant. For scientific method knows only an *object*, never a self-transcending, free, committed, and creative *subject*. When science through its method has spoken "officially" of man, therefore, it has found no shred of evidence of such a creative, autonomous self. It finds only a complex, natural organism conditioned in all it does by the various factors: genetic, physical, chemical, biological, psychological, and social, which have made it what it is and which for objective inquiry determine its subsequent career. The reality and effectiveness of human creative autonomy, of human subjectivity, has been vividly experienced, and in so far *known*, by the scientific community, and through them is indelibly impressed on human history. Yet that community when it employs the method that gave it this experienced and known freedom, knows but yet can know no such autonomy. Clearly, what this strange contradiction within science as a historical force signifies is that the reality which is experienced and known by the scientific community itself in doing science is much wider than the "reality" which the objectifying net of the scientific method itself can capture. This wider reality of the self, presupposed in science as a human activity, is known by the self-awareness of the scientist, his inner consciousness of himself as the subject of the knowing process: as held to the truth by his own commitment, as manipulating intentionally and freely his perceived world, as formulating new hypotheses and charting in reflection their future consequences, as testing them for a validity that is conditioned *only* by their congruence with his predictions, as being aware of himself as knower, and, as a consequence, envisioning new ways of shaping his world. Science *knows* the mysterious depths and freedom of the subject through the self-experience of the scientist. It is this self-awareness of the scientist as a committed, transcendent, free, intentional, and creative self that has given the sense of subjectivity, autonomy, and freedom to our recent cultural life. Clearly, science knows much more than officially it says it knows; that is to say, by self-awareness science *knows* the wise, potent man in the white coat, capable of informed judgments and new hypotheses, who writes the book, just as surely as it knows the object by inquiry, the passive and conditioned patient on the examining table about whom the book is written. An absolutized scientific method misunderstands the scientist who uses this method, and thus contradicts itself. This contradiction reveals the error of regarding this method as the *one* entrance into reality and truth; it shows how one-sided and, in fact, untrue to its own deeper knowledge a culture dominated by scientific method can be; and it helps to explain the deep and even angry reaction against an absolutized science presently so characteristic of our cultural life.

Second, an absolutized scientific method misunderstands the *object* of inquiry, that which the scientist seeks to know, insofar as what he knows through scientific inquiry is taken to represent the full reality of the object known. What cannot be known at all is not *there* for us; we use the word "knowledge" to specify what is taken to be real, independent, dependable, and in that sense "objective." To confine knowledge to one method — to a method that abstracts away from all subjectivity, centeredness, and uniqueness — is to constrict the world that is infinitely real to us in its depth and mystery; it is to objectify into a determined, subjectless realm all that with which we have to do. With men and women this is obviously a dangerous error, as if all *others* than the inquirers themselves were mere conditioned objects — empty spaces, as Tillich once put it — through which external forces pass. Such a view of human reality, expanded into the social and political arenas, would strip society of persons and create a social world of usable objects. We have pointed out the contradiction of this view of man with science itself as a creative human activity, as with the rest of life.

Such a view of the object of knowledge, however, also has devastating consequences for our relation with nature, as we are fast discovering. The relation of scientific method to the technological and industrial use, misuse, and ultimate despoilation of nature is not merely one in which technology applies, for the purposes of manipulation, the knowledge gained by scientific inquiry. It is also an *essential* relation consequent on the objectifying character of inquiry itself (4). Scientific inquiry knows by manipulating its object, by converting its qualitative Gestalten into homogeneous and so universal units (5), by investigating it with regard to its invariant relations with all that conditions it. Thus does inquiry strip its object of all its qualitative characteristics, its inherent integrity, unity, and centeredness. Such a world is known only through our manipulation of it; consequently, such a world has *reality* for us only insofar as it can be used by us for our own purposes. Objective inquiry, taken as our *sole* cognitive relation to reality, becomes the ideology of technological and industrial manipulation. To live creatively within nature, as well as with each other, we must allow ourselves to *know* the objects we encounter through participation and union, as well as through objectification and manipulation. Again, a mode of knowing that is creative as one aspect of our encounter with reality becomes destructive of nature and of ourselves if it is made absolute.

Third, as we noted, perhaps the major component of the

reverence for science in modern culture has been the promise that applied science has seemed to offer for human security and well-being. Our culture has swallowed Bacon whole: empirical knowledge, we have believed, is power to control the forces that run our world, the forces of nature, our genetic and psychological structures, and the forms of our society. Through such knowledge and the control it brings we can make our life infinitely better. History, however, has rudely wakened us from this Enlightenment dream, for control of the earth through technology has meant the misuse, pollution, and despoilation of the earth. Further, it has unleashed industrial expansion and appropriation that threatens soon to divest the earth of its available resources. Thus, far from guaranteeing human survival and well-being, the unimpeded expansion of our own technological control now precisely threatens that very survival and well-being. Ironically, the same cultural force that once promised to free us in the future from all that menaced us: disease, hunger, cold, poverty, and irrational tyranny — to free us from the "fates" — now is disclosed as *itself* a menacing and even mortal fate. History, and our own wills which help to shape history, now appear as much more mysterious, even demonic, than they once did.

The current despair about the scientific future to which I here refer, what has been called the Doom Boom, may well have been overworked, as are most prophecies, religious or scientific. But that technological development could produce such a "boom" and appear as a threat rather than a promise to our future bespeaks a vast change of cultural consciousness, a realization of the essential *ambiguity* of applied science which is quite new in our post-Enlightenment world. Homo Faber has been regarded by us as the creature who through his practical intelligence was of all earth's creatures most capable of adaptability. Equipped with modern technology, this same Homo Faber thus seemed the very paradigm of survival and of increased well-being. It now appears that Homo Faber can, through that same power of practical intelligence, destroy his world and himself with it. Does this mean that technological freedom, the power to shape and control events by intelligence, is the key to our *extinction* in history — our "fatal flaw" — and not to *survival*, as we thought? That would indeed be an ironic end to a culture that gloried and felt secure in that freedom! It seems this may be the case — unless man is *more* than technological man and learns not more about how to control nature but more about the control of himself. The absolutization of applied science as the cure of our problems, as the key to freedom from fate, has proved to be a mortally dangerous error. As the Greek and the Christian traditions have emphasized, more than technical knowledge is necessary for life; in fact, *techne* by itself, as our ecological crisis shows, is inherently self-destructive. Knowledge and control of the *self*, of its limits, of the infinity of its concupiscence, of its inherent waywardness and capacity for self-destruction, is also necessary, lest increased technical power spell disaster. This wisdom, enshrined in the mythical and religious traditions of our past (6), was regarded by a scientific and technological culture as a function of the scientific and technical weakness of early man or as a gloomy fairy tale made up by priests. It has, by the contradictory career of that same culture, now been empirically proved to the hilt. Here, I believe, is the deepest reason for the fall of the queen: the salvation that in the period of her pride she promised has turned out to be lethal.

Clearly, modern scientific culture had placed the problem of human existence at the wrong point. It had seen the major problems of life as stemming from our lack of control over the forces that impinge on us from the outside; thus, reasonably, it concluded that increased control over those forces would create an existence free of massive suffering and want. It forgot the mystery and ambiguity of the controlling self, of the *user* of science and technology, whose greater powers through knowledge may free him from external forces but who remains bound by his own greed and insecurity to misuse those powers and so in the end to destroy himself. It is the bondage of our will, not our ignorance or lack of power, that threatens our historical existence as a race. That science and technology could not by themselves cope with this more intractable problem is no fault of theirs; neither one is equipped so to do. But that in their day of glory they taught us to ignore these deeper issues and even to laugh them out of court *is* their fault. Again it was in the absolutization of scientific *techne* as a saving force in history that the error lay.

This point, that the problem lies in the self, in the will, in man himself, and not just in his intellectual ignorance or his lack of practical know-how, is not only an important lesson for the scientific community, it can also provide a defense in their behalf in a culture that is increasingly disillusioned with and distrustful of science. As with religion in the eighteenth century, many are now seeking to blame science for all our technological and ecological ills, to cry out for its rigid control or even extermination, and to regard it — much as free-thinkers once did religion — as a dangerous cancer in the body social. The fault, let me reiterate, lies not with science as intelligent inquiry nor with technology as the application of knowledge. These are in themselves good, evidence of the vast creativity of human existence, and replete with immense potentialities for human good. Incidentally, even though, as noted, they may now threaten our survival, they are also, paradoxically, utterly necessary for that survival. We cannot now do without them even if foolishly we would. What has been at fault is not our knowledge but our pride in our knowledge, not our technical power but our misuse of that power in the service of our material insecurity, our national pride, our insatiable greed. To put it theologically, it is sin not knowledge that now threatens us — just as before it was sin not ignorance that caused the most destructive of the earlier problems of our race. Like everything else that is human — including religion — science and technology can be, and have been, misused. To say this is to defend them from their fanatical detractors. But, let us note, to say this is *also* to admit that they are not omnicompetent, unambiguous, saving forces in history. Other ways of resolving other types of problems, other forms of knowing, other disciplines are *also* necessary in human existence, necessary precisely if science and technology are not to destroy us. The queen can save herself from banishment only if — as religion had to do — she is willing to abdicate her role as queen.

What is called for, therefore, is a reassessment of science in our cultural life, one conducted soberly by the scientific community itself and not alone by those who now distrust it. Such a reassessment is always painful, as it surely was for religion when it came under vigorous criticism in the Enlightenment and post-Enlightenment worlds. It was hard for those deeply concerned with religion and conscious of its saving power in themselves to face the fact that sincere piety could as a cultural force be a confining rather than a freeing factor and could be destructive rather than creative in social life. So it is now difficult for much of the scientific community and its lay adherents to give up the belief that scientific inquiry represents the one "pure," disinterested, and thus objective form of knowing, and the consequent faith that the application of organized intelligence to life's problem is likewise pure, disinterested, and unambiguous. The strange, contradictory career of a scientific culture, like that of a religious one, has proved otherwise. And as the free-thinkers of the Enlightenment foresaw the death of religion, or its confinement to special, deviant groups, so today as an example, Robert Heilbroner, himself a social scientist, foresees in the future a return to a static, traditional culture in which scientific inquiry will play a controlled, subservient, and very minor role. As an absolutized religion had proved destructive of social peace, so, he argues, an absolutized science and technology have become destructive of nature, of her resources, and thus of our common chance of survival. The only answer, he concludes, will be a return to spiritual and political levels of authority that will make the scientific and technological freedoms we now enjoy mere memories (7).

There is enough truth in Heilbroner's diagnosis — as there was in the antireligious diatribes of the eighteenth and nineteenth centuries — to compel the scientific community to rethink its role in cultural life. In a similar situation the only way religion has been able to recover its own integrity and rediscover a creative role has been to take herself through such a painful process of reassessment, yes, even of repentance and disavowal. When *others* reassessed the role of religion, she found herself banished from the court. When the religious community *itself* asked the question: If religion be neither the spiritual authority governing all of public life nor the queen of all the disciplines, what *is* her role and status, then a creative or at least tolerable answer appeared. Such is, it seems to me, one of the tasks of the scientific community as it faces what is surely to be a quite new future.

Although this reassessment is primarily a task for the scientific community, perhaps a friendly observer may make three suggestions. First, one of the "myths" of an absolutized science was that its knowledge, that is, the forms of its concepts and symbols, was purely *self*-generated, arising solely from scientific experience and scientific logic and thus in no way relative to the general notions circulating in its wider cultural environment. As orthodox religion, founded alone on divine revelation, disavowed any cultural and thus historical relativity to its dogmas, so an "orthodox" science saw itself as a purely cumulative discipline based only on its own inquiries. Scientists *enjoyed* the belief that although the content of other disciplines depended on what science knows, scientific knowledge was in no way dependent on the knowledge gained in other disciplines.

Recent studies in the history of science, however, have shown that this self-understanding was in great part an illusion (8). The categories, models, and paradigms of scientific understanding, at each stage of its development, have been related to and in many cases directly dependent on notions generated in other fields. As a human and thus cultural activity, science is *relative* to its historical context, expressing through its own specialized categories many of the economic, political, social, psychological, philosophical, and even religious presuppositions that have determined that environment — as, of course, does any formulation of religion. It does not, therefore, represent a "pure" or totally objective form of knowing validly dominant over all the relativities and partialities of culture in its other aspects. At each stage of its life science itself represents an aspect of that same relative cultural vision, itself therefore to be corrected and supplemented by other aspects of culture and even by other cultures — and not only by its own future developments.

It is well known that no form of "orthodoxy" — orthodox Catholicism and Marxism have been good examples — understands its own doctrines through a careful study of the *history* of their development. On the contrary, each orthodoxy understands its own history only in the terms of its present and absolute dogmas. It is not insignificant in this respect that science is *not* taught to young scientists in terms of its history. In the vast majority of cases science is studied and so viewed by young scientists only in terms of its present point of development — as if controlled experiment and logic provided its only components and the cultural matrix out of which it arose were irrelevant to its full understanding. One suggestion, therefore, in the reassessment of science is that it be studied historically, as well as systematically — as, at present, are art, social theory, literature, philosophy, and religion — with the express aim, as in those other disciplines, of showing the intrinsic relation of its major models and paradigms to their changing cultural contexts. As religious orthodoxy has found, history is an effective detergent of absolutization. A historical view of science will help in the achieving of a realistic reassessment of her actual role in cultural life.

Second, as we have argued, scientific inquiry does not represent the sole cognitive relation men and women possess with what is actual. As Whitehead argued, like the sensory experience on which it is based, scientific method abstracts for certain purposes from the totally encountered world, from our constitutive relations with things, from awareness of the subject of knowing, awareness of natural beings around us, and awareness of persons as persons. To confine knowing to this one significant but objectifying method is to strip natural objects of their inherent reality and value and persons of their selfhood, their creative freedom and thus their humanity. Science must, therefore, see itself as only *one* aspect — to be sure, a most important and valuable aspect — of human cognitive creativity, and thus one supplemented by and dependent upon other aspects if it is to take its rightful and not dominating role in our cultural life. Such a reassessment implies an acquaintance with other modes of cognition: in literature, in social existence, in the arts, and in religion. And it necessitates an understanding of the interrelations of these modes of knowing to science that only philosophy of science and philosophical epistemology can bring (9). When theology lost its absolute base in revealed dogma, it was incumbent upon it to reinterpret religious knowledge in relation to the other valid modes of knowledge in culture, to history, to natural science, to psychology. Thus did the discipline of philosophy of religion come to prominence; such a critical and philosophical interpretation of religious knowledge is now a part of all advanced education in religion, as is a study of religion's history. A corresponding reassessment of scientific knowledge in relation to the other cultural modes of encountering, knowing, and shaping reality would set science among the humane arts and thus help to humanize rather than to dehumanize our common world.

Finally, as we have seen, the application of scientific knowledge has revealed itself to be an instrument of man's will and thus subject to all the distortions of which that will is capable. Knowledge is power, and power can corrupt, even when that power springs from knowledge gained through objective inquiry. Informed intelligence can be the servant of our greed and our desire for security; it has not, as was hoped, been their master. Apparently, the more technical know-how we possess, the freer we are to ravish the earth and to plunge ourselves into a new unfreedom of scarcity, conflict, and ultimate authoritarian control. It has been another cherished myth of our culture that technology raises only technical problems and that to every technical problem there is in potentiality a technological answer. Thus again through myths about themselves science and technology were regarded as self-sufficient in our social life, dependent on no other aspects of culture for their own self-realization and for the realization of greater well-being for all. These myths — the developments of medical expertise, genetic capabilities, urban culture and, finally, the ecological crisis — have exploded. We now know that technology raises political, social, legal, and moral problems on every front, and consequently that the proliferation and expansion of our technological capacities must be guided by legal, social, and ethical wisdom if self-destruction — or sheer political authoritarianism — are not to result. The increase of man's power through applied knowledge has not increased his virtue or his wisdom. Rather, by threatening his human well-being if not his survival, that increase has raised the question of his virtue and his wisdom more sharply than ever. A scientific and technological culture has not made the existential and moral questions of religion irrelevant, as it had thought. It has posed those questions anew, and in a terrifyingly intense form, as crucial to the realization of that culture's own potentialities and to avoidance of that culture's own self-destruction.

Unfortunately, as Socrates and St. Paul both knew, there is no available educational program that can guarantee either wisdom or virtue to any of us. Possibly, as in current religious studies, an acquaintance with the sociology of science — how ideological, class, national, and professional prejudices affect the application of scientific knowledge to the world's problems — and a study of the ethics of science will help a little. Together, sociology of science and ethics can, if taken seriously as a part of the education of the scientist, open the eyes of the future scientific community to the dangers of their own increasing power, to the responsibilities of their future role, and to their intellectual and moral dependence on legal, ethical, and religious wisdom if they are to be creative and not destructive in our common future. In the end, however, science and applied science — like every other aspect of human creativity — must learn to live and deal with the vast ambiguity of *their own* creativity — which is an existential, moral, and religious problem facing *every* profession, but new, I suspect, to the scientist as it once was new to the priest. For the lesson of history, and now of the history of a scientific culture — and surely also the message of the gospel — is that it is the very creativity of man that can spell his doom, that his knowledge can be turned into blindness and his power into self-destruction. To recognize this mystery latent within even that which is most creative in our life is a part of wisdom and the beginning of repentance. Such repentance on the part of all of us who have helped to create the very dubious destiny we shall bequeath to our children will possibly help to soften that destiny. Without such repentance and such new humility by a scientific and technological culture, the future that science brings to us — as well as the future of science itself — may well be darkness and not light.

Such wisdom and repentance may also become an entrance into a much deeper faith. As we have seen, modern culture experienced and had confidence in the promise of life and of the future because it believed that informed and organized intelligence, to use Dewey's favorite phrase, could resolve our problems. Now that such intelligence, both as science and as technology, has revealed itself as essentially *ambiguous,* as raising as many serious problems of survival and of well-being as it solves, the question of the meaning of our history — a philosophical and a religious question — is again forced directly on us now by the career of a scientific and technological culture itself. Not only do the two of them reveal the reality of the bondage of our will, they disclose as well and anew the ambiguity and deep mystery of human history, the real possibility of self-generated catastrophy, and thus the need for a deeper basis for our confidence in the future than our own virtue and wisdom. If we look carefully at the ecological evidence, the future of our scientific civilization can be presented in dark colors indeed. Despair and not confidence seems in truth to be the issue of a technological culture when it has run its full course.

A word, however, must be said in as sharp contradistinction to that new despair as to that culture's former optimism. Our human history is not compounded merely of human creativity and of our destructive use of that creativity, both of which science and technology have now disclosed to us. There is also the Lord who brings judgment on cultures that are too proud of their wisdom and power, who gives the possibility of repentance and of new life to those who listen to that judgment, and who — as with a captive Israel — always holds out the promise of a new covenant, a new act, a new possibility in history that may redeem the times and bring light even to the future that is coming.

REFERENCES

1. For example, on the day this page was written (Aug. 24, 1975) *The London Times* carried the announcement by the American Institute of Physics and the University of California at Berkeley of the possible observation of "a particle representing the basic unit of magnetism" which "could rank as one of the major scientific events of the century." Immediately after specifying the discovery, the announcement went on to say, "If the particle could be captured and controlled, some of its practical uses in medicine and industry could be revolutionary," and it proceeded to outline some of these "exciting" uses in medical therapy, in providing new energy resources, in more efficient motors and generators, and so forth and so forth.

2. The references to Dewey's philosophy are taken from the Modern Library edition of Dewey's works, *Intelligence in the Modern World, John Dewey's Philosophy*, edited by J. Ratner, Macmillan, 1939. For the unity of science and technology, see pp. 315–317, 320–325, 327–334, 757, 945–948.

3. Ibid., pp. 294–297, 357–358, 759–760.

4. Herbert Marcuse, *One-Dimensional Man*, Sphere Books, 1968, Chap. 6.

5. Ratner, op. cit., pp. 324, 337–340.

6. The correspondence of this history of technological capacity leading to a self-inflicted doom with the Prometheus myth is a most fascinating and chilling aspect of the present situation. In this respect, for all its technical inadequacies, early Greek culture seems much wiser and more sophisticated about the ambiguities of human creativity than does our own recent culture.

7. Robert S. Heilbroner, *An Inquiry into the Human Prospect*, Norton, 1974.

8. Thomas S. Kuhn's work (*The Structure of Scientific Revolutions*, Princeton) is of course the most renowned in this area. But research by John M. Greene, Stephen Toulmin, Charles Gillespie, and Herbert Butterfield, not to mention the classic work of E. A. Burtt (*The Metaphysical Foundations of Modern Empirical Science*) has led to the same conclusion. Science, even at the highest levels of its theoretical development is *relative* to its cultural context and not an independent, self-sufficient, or autonomous body of knowledge.

9. The gulf between philosophers of science and working scientists is to a lay observer amazingly great. Most scientists seem quite unaware of the important and diverse discussions of the logic and epistemology of the scientific method, and its relation to knowledge in other fields, conducted by philosophers of science. As one brilliant physicist said when he visited a seminar on religion and science and there heard his university colleague in the philosophy of science discourse on the "problems" and "puzzles" involved in the understanding of the scientific method, "I don't know anything of or pay any attention to these problems. I just go into the laboratory and conduct my investigations; and these questions whether, what or how I know what I know are neither real nor relevant questions to me."

    Such an answer shows, to be sure, a healthy state of professional self-affirmation. But when I translated it back into the terms of my own discipline, he was horrified: The minister, hearing that an educated layman (a chemist) had "problems" in relating religious knowledge and certainty to his knowledge of science and of its method, says to that layman: "I don't know anything about those 'problems' you mention, nor do I pay any attention to them. I just go into the church and pray and preach. What I know is there, how I know it, and whether I know it are 'layman's' questions; they spring from unfamiliarity with the doctrines of our faith and thus are neither real nor relevant to me or to the community of faith. Only believe, and you will be saved."

    The parallel is of course by no means exact, but it may be informative. Both scientist and minister are "professionals," and so they have both "boxed off" their professional worlds from their own actual lives as members of a wider culture and as participants in a much more complex existence. Each seeks in his own way to avoid the difficult and often painful effort of putting their simplified professional world together with the very complex *actual* world. But the layman can't do this; he must bring the worlds in which his life participates together into one — and thus in his "simple" way he asks the most profound questions, the questions that, like a child's questions, probe beneath the walled-in worlds of specialization to the mystery or their unity. A child's questions are not "childish." They can be answered only on the level of philosophical and theological formulation.

Perhaps the largest difference is that the minister would, unlike the scientist, be regarded as a *poor* representative of his profession in giving this answer. Not to take seriously the questions of the layman about religion is, for the modern mind, to blind oneself to *real* questions, important questions, questions that should and must be answered by any religion that is alive rather than dead. On the contrary, the scientist has no such attitude to the questions the layman may put to him about science: about the status of its truth, how that truth relates to other forms of truth, the status of the reality known in science, etc. He regards these questions as naive, "simple" or "strange," showing little understanding of science (which knows they are irrelevant or meaningless, and which "does not need to ask them" — as if *this* were relevant!). Thus to the scientist these sorts of questions appear as functions of the *ignorance* of the questioner, his nonmembership in the scientific community, and so of little concern for the working scientist. Significantly, he regards science as "belonging to him" or, better, to the community of scientists, since they do understand it — as priests once regarded religion. This is, of course, a mistake, a professional prejudice. Science is as much a part of the life of the layman as is religion — and as it is a part of the extra-professional life of the scientist. Thus the layman *must* deal with it and with its view of reality in relation to other aspects of his life. Thus he *feels* the problem of scientific knowledge in relation to other forms of knowing, as he feels the problem of religious knowledge in relation to other forms of cognition. And the questions in each case about the reality there known are no idle or silly questions but utterly fundamental, for the scientist and for his community as for the layman. Science will be internally healthier and externally more respected, as is religion, when *its* "layman's questions" are listened to and pondered, not just by laymen but also, and this may be new, by scientists. The scientist, like the minister, may well then find that the puzzles of the worried layman and the questions he asks are the ones he is perhaps least capable of answering — and so where, in order to understand himself and what he does, he may need the help of the philosopher.

DISCUSSION

General Comments

Dr. Weller.   It's difficult to sort out one's thoughts in response to Dr. Gilkey's lecture. He depicts a crisis situation in which science is set up as a straw man that should be burned in effigy as being responsible for all the faults of human nature. I am concerned that the approach is somewhat parochial. It is based on the Judeo-Christian understanding of ethics and doesn't look at the world as a whole. It is assumed that there have been two recent queens — one displaced and the other one in the process of being displaced. I'm not sure that science is a queen; I'm not sure that it's dominant, and I'm not sure that it will be displaced. But if it is, if we accept these things, how do we project our problems today in terms of the world as a whole, and what is the responsibility of the two named queens for the problems that we face?

When I started teaching in about 1940, there were some 2 billion of us on this earth. Now there are 4 billion. What is the relative responsibility of the two queens for this increase in population? Both queens have had their role. Currently we add each year the number of people that would populate a Bangladesh or a new Central American country. This is producing pressures on a global society that we have never experienced before. However, until technology and science solve some of the problems of society on a global basis, face it — we can't expect to bring the disproportionate death and birth rates into some sort of balance. There will be new problems to face,

---

Editor's note. Asterisks denote a break in sequence or a change from one discussion panel to another.

and unless society has faith that science will be able to meet this continuum of new problems, I think we're going to be in a very bad way.

*   *   *

Dr. Lamb.   I was very impressed with Dr. Gilkey's fluent and forceful presentation, but I think I detected a little bit of nostalgia for the good old days when things were more simple. I'm ﹅not very well equipped to deal with historical remarks, but I think that a few centuries ago in London the smoke from coal fires was so great that at one point the King ruled it would be a capital offense for anyone to use coal unless he gave them permission. This kind of pollution took place long before science could be blamed for it.

The water underneath the city of Tucson is sufficient for the present, but I'm a little relieved that I won't have to live there too many decades longer, because I think the water will begin to taste quite salty. Now, I don't really feel much guilt as a scientist for the problems of pollution and the use of natural resources. I think the problem is that society makes decisions about what will be allowed. I have attended a few meetings of planning commissions where more building was authorized than should have been. It also seems to me that Congress is under very great pressures to permit all kinds of activities that lead to destruction of natural amenities. I happen to believe in democracy, but I think there are difficulties that arise from that system. The earth has a limited surface area and a limited amount of natural resources, and we just will have to learn to somehow live with it.

*   *   *

Dr. Schwinger.   I believe that science was compared to a priesthood, and I would like to hope that science will never become an isolated priesthood, but, in fact, that the knowledge of the science will be shared by an educated and informed public, and that science will never have this secretive aspect that the priesthood insisted on. In fact, a proper appreciation of science and its fundamental role in guiding and in enlivening modern life can only be carried out if it is properly understood and supported by an intelligent public.

Dr. Kuznets. Historical analogies are easily drawn and easily overdrawn. Dr. Gilkey's presentation, in a sense, drew a complete parallel between the role of the priesthood, organized religion, and theology in the centuries in which religion was dominant in European Middle Ages and the position of the scientist in modern society. I frankly don't think the analogy holds. It doesn't hold because the scientists have never been given the power within society that was claimed, and for some centuries actually held, by the religious authority. Neither did the scientists, as far as I know, claim an intellectual predominance or exclusiveness, which would bar any other forms of knowledge and enforce it by physical means. So I think we ought to be very careful.

Now to be sure, the technological power of mankind for both harm and good has been magnified but I would like to conclude by suggesting that by any calculation, the losses resulting from technology in very simple terms, in terms of human lives lost even during world wars, have so far been a very small fraction of total population — as compared with the kind of losses that were sustained with a more primitive technology during the Thirty Years' War. We are forgetting that despite the horrors of technological power and allowing for the danger of an atomic holocaust, so far the negative impacts in terms of losses of human lives and even in terms of ecology, have been relatively slight when you view them in a long-term historical perspective.

Dr. Eccles. I would say a mistake was made in identifying science and technology. This was a mistake for the purpose of building a case. The case couldn't be built on science alone. He had to put in technology with it, say it's the same case, and then we stand convicted. I'm not going to be convicted on those grounds.

Dr. Anfinsen.   I don't care to be convicted on the basis of the presented evidence either. As a matter of fact, I don't believe that in private conversations Dr. Gilkey would take the kind of stand he took. What I would suggest would be that men like Dr. Gilkey and men like those of us on this platform get together. He represents the image of security and safety to a lot of people in this population, and we do to a lot of others. The people we have to attack are, in many cases, the media, the corporations, the military, and the politicians. We should do this as partners rather than as argumentative people at a conference.

Dr. Dean.   I have great sympathy for the men on this panel. I've been in the position many times myself. I've been accused of all the faults of religion and I've taken it personally and felt offended that I was blamed for the Thirty Years' War and the Crusades. And I think sometimes I might have been so ready to be offended that I misunderstood the person I was talking to. I think if we listen more closely to what Dr. Gilkey was saying, he wasn't saying that scientists are personally responsible for these things. If you look at his language closely — and I could quote from the fourth page — generally he's speaking about what has happened to science and what has happened to it as a result of what this culture has done to it.

He says that science has been a religious force for our culture. He says through that "sacred" knowledge there is given to those who bear it the status of priesthood. I think that the thrust of Gilkey's argument was not to call into question your morality, and I don't think it's appropriate to respond as though he did. I think he's saying there's a cultural condition that's arisen. And that's the condition wherein science is given a certain sacrality. Scientists have been given a certain priestly status.

They may be as uncomfortable with it as Dr. Gilkey is, but that may be the fact regardless. I think all of us would know something of what that's like. I think it would be more appropriate to address the cultural situation in which science has been given power and maybe a religious status by the culture, and see whether that's proper.

* * *

Dr. Harvey. These meetings depress me, frankly. The reason they depress me is that I don't think we listen to one another, and it's a commentary for me on academia generally and intellectuals in particular. I don't see how anybody could have listened to Dr. Gilkey's address and drawn some of the conclusions that have been drawn here from what he said. On the other hand, I have no wish to identify myself on the whole with the way he did it. I don't think we theologians are in any position now to sit back and look at a new queen and ask them to "put their house in order." Religion, in my view, has a very sorry history in the Western world. I am particularly sorry about the exclusivist claims of Christianity which seem to me to have led necessarily to anti-Semitism and to a certain kind of racism. I happen to think this is endemic in Christianity, and, therefore, I don't feel that as a theologian we ought to be in the position of criticizing science.

The other side of the story, though, is that as human beings and intellectuals we ought to ask one another, and particularly we ought to ask scientists what we can do as intellectuals and as academics about illuminating and doing something about the problems we have in the world. One thing Langdon's address made clear, and I don't see how anybody in his right mind can deny it, is that science has become the cultural force that he said it is. I don't see how anybody can live in a modern university or in this culture of ours and deny that. I haven't any interest in creating categories like queens and asking them to repent, but I certainly think I have a right to ask the scientific community to reappraise and to take the lead in reassessing what science means in our culture, what it means in the university, and what it means with respect to problems in the world.

*   *   *

Dr. Kusch.   I'm a little confused about modes of cognition other than a sort of intellectual cognition. Dr. Gilkey seemed to place a great deal of emphasis on alternate modes of cognition. It is true that if I look at a painting I obviously don't subject it to scientific analysis by looking at the spectrum of the reflected light. I don't do that. Nevertheless, I find it hard to think of alternate modes of cognition as powerful as the modes of cognition of science. I don't think you can substitute any other mode of cognition, and I recognize it exists. My pleasure in listening to the Brahms Concerto has, I admit, a place, but I know of no other mode of cognition than that used by science which will allow man to deal effectively with the world in which he lives.

Dr. Cobb.   I do think it's important to consider what we might mean by alternate modes of cognition a little bit more directly than that. Just to give an example, let's consider the whole problem of cognition of our historical past. We have large segments of universities devoted to this kind of inquiry, and one might say this is less exact, and certainly that it does not lead to the same kinds of capacities of control that scientific inquiry does. There have even been some who think that historians can only be truly disciplined and responsible, and can only be genuinely critical insofar as they model their methods upon the methods of the sciences.

   It does seem that the methods of history whereby we reconstruct our past, and that also in some ways give us clues as to the future, are a part of our way of understanding who we are and where we are, and that this is different from the model of scientific knowing that has in fact been most dominant in our culture in the past. The problem of a queen is that her method is used not only in the discipline where it is unquestionably the best possible method to be used, but also is emulated in disciplines where that may not be the case.

The same thing may be true with certain aspects of psychology. Some people, at least, feel that the efforts of the study of the human person, emotions, and interpersonal relations have been limited in ways that were unnecessary by the attempt to treat human beings as if they were like other objects of cognition in an effort to assimilate psychology to the natural sciences. There may be other ways to understand the human situation which are highly disciplined and which are not therefore just inferior imitations. I think that's the kind of thing he had in mind by multiple modes of cognition, which should be allowed to freely develop and be responsible in their own terms rather than be subordinated to one pattern.

Dr. Mulliken. I think that what Dr. Cobb said about the scientific method having been set up too much as an ideal in the social sciences is very true. In the social sciences there has been a regrettable excess at times in trying to apply the methods of the more quantitative sciences where they should not be applied. But I believe the scientific attitude and the scientific method are just as applicable to psychology or to historical analysis, if you like, but with the proper appreciation of the very different degree of uncertainty which exists in these different fields.

* * *

Dr. Schwarz. I think there is the point where I wondered whether the image that was mentioned here, the two queens, is not too simple in the whole context. I think very decidedly you said, and I agree with you, that we are at a point of change where we have to reevaluate the business of science as much as we have reevaluated the business of theology. I think an issue that I saw in your paper, and perhaps should have drawn out more decisively, is that this is the way human nature works. What we have done is put science in an idol spot. This is the god that we worship. Of course, since an idol cannot produce the way God can, we now wonder about the results and blame the wrong people. Instead of blaming the scientists, we should blame ourselves. What have we done?

Dr. Gilkey. I would first of all like to say that perhaps the main theme of my point is that human nature is the problem. I agree entirely at that point. If that is so, as I suggested in one paragraph, it seems to me to be both the logical and effective defense of science and technology against its more stringent detractors, and there are a lot more in numbers than myself. But it also means that it's not omnicompetent, and I think this an important point. Human nature is the problem, there's no doubt about that, and we all are involved in this.

I would like to stress that I think this is perhaps a point of difference in the way we analyze things and that such an analysis of mine is not directed at individuals. I don't like to overwork the analogy, but as one begins to talk about the problems of medieval culture with religion or sixteenth and seventeenth century culture with religion, one is not directing one's point at the individuals involved in Lutheranism, Calvinism, and Catholicism, in particular. I think the social scientist and the historians have shown us that there are intelligible entities to talk about that one can call communities — social forces, historical forces, cultural epoch, points of view, etc. It's that, that I'm talking about being, just as I think it's quite relevant to talk about the problem of being American in the world without particularly wishing to speak of American individuals. Certainly there is a way in which we all participate in the responsibility involved there, but it's not, however, an accusation of individuals for insincerity or arrogance.

*   *   *

Question.   Is it the judgement of the panel that science has lost significant prestige in recent times, and if so, why?

Dr. Bloch.   I think the answer to the first question is yes. It has lost prestige. This is a sheer matter of observation. I've talked to many of the young people, students, my own children, and I've checked with other people as well. Generally speaking, especially among the young people, the faith and admiration for science which I think I sensed much more 10 or 15 years ago has been strongly diminished. So I think the answer to the first question is yes.

The answer to the second question, why this has occurred, is more difficult. Of course, it's natural for scientists to say it's not my fault. And to some extent this is certainly true. However, one cannot say either that there is no reason for this disillusionment among the young people toward science. Of course, you can say to them, you must distinguish between technology and applied science, but that is difficult for them. Maybe one should try more to explain that difference.

But I believe there is a deeper reason to it. That is to say that they feel uncomfortable with a world view which is purely rational. I think we now come to the problem that Sir John raised today. It is true, if you imagine a future world as a type of brave new world of Aldous Huxley where everything is regulated and everything is done rationally, that such a world is a horror and that's the way Huxley meant it. I think the only thing one can hope for, is to point out just what Sir John and I think others of us have kept on emphasizing — that we do not claim and we could not possibly claim, that pure scientific reason is the image of the world which mankind should thrive on; that we are sensitive, and in fact perhaps more sensitive than other people to the emotional, to the poetic, to the religious, if you will, aspects of human life. I believe there may be a meeting point.

Dr. von Euler.   I think I would like to go on a little on the theme which Dr. Bloch started. It seems to me when talking to people in general, family and friends, and so on, that many people are afraid of science in a way. There is something in science now, a connotation, which frightens them. I think that's partly due to the fact that they have not had any particular training in science, they don't know enough about it. And if one then would look into the possible causes of this, I would like to suggest that the mass media do not always treat science in such a way that it induces confidence in scientists.

*   *   *

Dr. Bethe.   I think people do fear science and scientists, and I believe they feared them even when science was held in very high esteem just after World War II. Certainly they were always in awe of science. Now science is being attacked by many people, and the address by Dr. Gilkey brings this out in more detail. I believe people are afraid of science because they don't understand it. They find it requires a very big mental effort. . Very often it is our fault because we don't explain it well enough, but also very often it is the fault of the public that they don't take the trouble and the mental effort to try to understand. We have to admit that science is a subject which you cannot understand in one day or in one hour's radio talk.

Responsibility and Science

Question.   Is the scientist personally responsible for his role as a culture-altering force?

Dr. Kusch.   Again, there seems to be a special mystique to the scientists, or there is believed to be. You, in the middle of the third row, play a culture-altering role by purchasing a television set. In fact, every human being, whatever he does from the most trivial to the most profound, contributes to the change or the development or decay of our culture.

What troubles me about much of this conversation and other conversations is that somehow or other what I do, and what Dr. Hofstadter does and what Dr. Mulliken does, is in a special category. You look at our day-to-day work, the things we do, as a sort of a unique culture-changing activity. It isn't so. The culture is changed by the building of this building. It had some modest cultural impact. I find it hard to think of anything that I do which doesn't somehow or other modify the culture. And I am perfectly willing to grant that what I have done as a scientist, and even more what I have done and said as a teacher, has been an immense, for one man, an immense cultural influence.

Dr. Cobb.   I fully agree with what's just been said, but I want to comment on one feature of it. I think that the fact that the scientists on the panel are all vigorously denying that science is queen is very interesting, and I'm glad they're doing it. Don't misunderstand that. The question, however, indicates the sociological fact that over a period of a couple of centuries science has been viewed as queen, so although, of course, all of us are engaged in culture-modifying activities, we think that most of what we do is trivial in comparison with the enormous cultural modifications that are effected by this particular community of people whom we admire, revere, respect, and are afraid of; and that's what's meant in this whole queen language, however objectionable that may be.

I think this question is coming out of the context of that attitude and I'm glad that the scientists on the panel are working so hard to demystify their own discipline. We need that. At the same time it would be a mistake to think that most of the rest of us have had, in fact, the culture-modifying effect that the distinguished scientists who are here gathered have had in their lifetimes.

Dr. Hofstadter.   Well, if that is true then I think that Mr. Gilkey should have told the audience to repent.

Dr. Cobb. Excuse me, sir. I really believe, in part, you did misunderstand. I think he is saying that scientists have participated to some extent in creating the mythos, the erroneous image. I don't believe that he intended to indicate that individual scientists were doing this anymore than the rest of us.

Dr. Hofstadter. Who's doing it?

Dr. Cobb. Well, when you're dealing with vast historical movements, such as a couple of centuries of Western history, it's very difficult to pick on individual people.

Dr. Hofstadter. Groups?

Dr. Cobb. Groups? I would say educators, philosophers, scientists, theologians, etc., and, above all, governments, military groups. It's a very, very widespread characteristic of our culture that we have looked to science as informing a technology that jointly would be able to solve our problems. We had hoped for salvation, and I did hear one member of the panel earlier at another session indicate that he believed that science was the salvation of mankind. So it's not something that scientists are completely free from. But I would not single out scientists as those who make the greatest claim for science.

Dr. Hofstadter. Yes, well, I still think that Dr. Gilkey should tell the audience to repent, and the media to repent, and almost all of society. Why pick out the scientists? There's nothing special about the scientists. They don't make the policy. They are the objects not the authors of this policy.

Dr. Kusch. I agree totally. Unfortunately, however, perhaps there is a misconceived belief in our public that we play a role which we don't in fact play. That is what some of these questions suggest. This is, I think, a reality, and I think we must cope with it in some way whether that perceived reality has a basis in fact or not.

* * *

Question. A lot of discoveries which seemed great have turned out to be destructive. Is this because scientists don't realize the impact their discoveries will have on society? Doesn't the scientist know that a new discovery is a form of power and that power is a very dangerous thing?

Dr. Kusch. There isn't a discovery that has ever been made which couldn't be directed to destructive purposes — although there may be a few exceptions. The question has implicit in it assertions which I would challenge. Let me go back to the matter of nuclear bombs. A couple of crazy Germans observed nuclear fission, a laboratory curiosity with primitive apparatus. There was no reason for them to foresee this as a weapon of the century, if you like. The ballistics which allow intercontinental missiles to fly were first described rather decently by Galileo, it's as old as that. But, in fact, Galileo made contributions to the flights of projectiles, not nuclear weapons.

It's a common attitude that in careless villainy we are destroying the world. If I knew that anything I wrote in a paper would have some sort of unique quality of destruction implicitly, I would tear the paper up and burn up the notebook. But life isn't that simple. It isn't predictable what use will be made of it. It's not within the capacity of any individual to do so.

Dr. Hofstadter. I agree with Dr. Kusch, but if you carry the question to its logical conclusion, you would say, why use logic? In fact, logic is the real instrument by which you do all those bad things, as well as all the good things, and you just can't turn off logic if you want to live in this world.

Dr. Kusch. I have said this about nuclear weapons. We wouldn't have had nuclear weapons if we hadn't had a thriving sewing machine industry in the United States — a certain kind of technical skill, a capacity to create sewing machines. I don't mean sewing machines specifically; I mean things of that kind. The surest way of not having damaging technology is to hang the teachers who teach your children to read. I mean this. The original sin is learning to read as much as anything.

Dr. Yang. I agree with Professor Kusch but I would like also to add this point. I think that science and technology are necessary for mankind. In many senses, development of science and technology is almost a natural phenomena. So we might as well live with it.

But it is important for the scientist or the engineer who engages in this, because of his deepened perception of what is going to come from his discovery (clearly, he has a better view of it than other people), to have a deepened feeling of responsibility toward society.

Now if one takes this viewpoint one can examine the record of the scientist, the physicist, the chemist, and the biologist in their activities. My conclusion is that, contrary to some popularly held views, the scientists have been collectively, in general, quite responsible. To give an example, the study of genetics in the biological sciences in recent years has produced a possibility of manipulating genetic material in very dangerous ways. I've read that the biologists collectively have discussed this question, and exhibited full recognition of the possible social implications of this matter.

I think one should make it a responsibility of scientists and engineers to warn the world about the possible dangerous social effects of their discoveries. I would also think it is the responsibility of the people at large not to say that these mad scientists are irresponsible, because I believe the record shows the reverse.

*   *   *

Question.   Should a scientist ever feel guilty about a discovery he's made?

Dr. Beadle.   The answer I would make is that scientific discoveries themselves are neutral, but that the use of such discoveries are usually not. And science as a whole, pure and basic, is neutral. Both have applications that can be for good or bad.

Dr. Gilkey.   I think in all these things there's no way to escape guilt in human existence one way or the other. I think it's good to recognize that. It's guilty to be an American, it's guilty to be a white person, it's guilty to be wealthy. I think any group that pretends that they are going to escape the problem of responsibility and guilt are going to have terrible troubles, if not personally, then in history. And I think this is true of religious groups.

I think it's also true of any of us who contribute to the knowledge of a culture in whatever way we contribute to it. This knowledge is going to be used, and we are in part responsible for it. However, having said that I think human life faces the problem of guilt, there is no position in life that is guiltless, and one has to deal with this. I think that to localize guilt in this case toward the scientists, or even the technologists, or the guys who ran the airplane, is absurd. Whatever there is, we all share in it. As Dr. Kusch pointed out in conversation just a few minutes ago, if you're going to blame the scientists who found out about the relation of energy and mass and so forth, I mean the pure scientists, then the guy who invented the wheelbarrow is guilty for the development of the culture that led to this. There is no escape from this.

Culture is an ambiguous thing. We participate in it. I think that if we like the roses that we get, we've got to take the thorns. I think the scientific community likes the roses, and I'm not sure one can have the roses without a prick of the thorn once in a while. The roses are the social acclaim, the social prestige, and so forth. And the thorns are there too. But I don't think that the scientific community is in any part in a special localized place of guilt in this business — at all. That's not the point anymore than that the preachers are entirely responsible for the excesses of religion.

Dr. Walton.   I'd like to speak to this question of the responsibility of the person who makes a new discovery. I would like to describe to you a situation which I experienced which may interest some of the audience here. The year is either late 1932 or early 1933 and I was a research student at Cambridge, England. Now it was a time when there was a very strong pacifist feeling amongst students at the universities, and a common matter for discussion was whether it was right for physicists, let us say, to take a job in a firm that was concerned with the manufacture of armaments. You have to remember those were the years of the great depression, and you were lucky if you could get any sort of job at all.

This was the time when the Oxford Union held a famous debate, and the motion before the house was that this house refuses to fight for King and country. Oxford Union decided in favor of this motion. This horrified the people of England. They thought that the coming generation was going to let the country down.

I can remember at the same time a discussion in the Cavendish Laboratory. Now at that time the Cavendish Laboratory was one of the foremost laboratories in the world for the study of what was called atomic physics. It was a discussion among the research students on this subject of responsibility. And I can remember very distinctly one student saying to the rest of us that he felt very thankful that the sort of research that we were all engaged on in the Cavendish Laboratory couldn't possibly be used for warlike purposes. Well we know that it didn't turn out just that way.

But I go a little bit further and quote to you something about Rutherford. Now Rutherford was at this time a world leader in this subject which we call nuclear physics. He died in 1937. Sir

George Thompson, son of J. J. Thompson, wrote an article about 15 years ago, in which he described how in the last year of Rutherford's life, on two separate occasions, he heard him express the opinion that we would never be able to get out of the nucleus, the energy which we knew was in the nucleus. And yet the atomic bombs went off within a few years of that. Now if a world leader in the subject of that sort cannot see a few years ahead, what is the ordinary mortal to do about foreseeing the future?

*   *   *